List of titles

Already published

Cell Differentiation	J. M. Ashworth
Biochemical Genetics	R. A. Woods
Functions of Biological Membranes	M. Davies
Cellular Development	D. Garrod
Brain Biochemistry	H. S. Bachelard
Immunochemistry	M. W. Steward
The Selectivity of Drugs	A. Albert
Biomechanics	R. McN. Alexander
Molecular Virology	T. H. Pennington, D. A. Ritchie
Hormone Action	A. Malkinson
Cellular Recognition	M. F. Greaves
Cytogenetics of Man and other Animals	A. McDermott
RNA Biosynthesis	R. H. Burdon
Protein Biosynthesis	A. E. Smith
Biological Energy Conservation	C. Jones
Control of Enzyme Activity	P. Cohen
Metabolic Regulation	R. Denton, C. I. Pogson
Plant Cytogenetics	D. M. Moore
Population Genetics	L. M. Cook
Insect Biochemistry	H. H. Rees
A Biochemical Approach to Nutrition	R. A. Freedland, S. Briggs
Enzyme Kinetics	P. C. Engel
Polysaccharide Shapes	D. A. Rees
Transport Phenomena in Plants	D. A. Baker
Cellular Degradative Processes	R. J. Dean
Human Evolution	B. A. Wood
Human Genetics	J. H. Edwards

In preparation

The Cell Cycle	S. Shall
Microbial Metabolism	H. Dalton, R. R. Eady
Bacterial Taxonomy	D. Jones, M. Goodfellow
Metal Ions in Biology	P. M. Harrison, R. Hoare
Muscle	R. M. Simmons
Xenobiotics	D. V. Parke
Biochemical Systematics	J. B. Harborne
Biochemical Pharmacology	B. A. Callingham
Biological Oscillations	A. Robertson
Photobiology	K. Poff
Membrane Assembly	J. Haslam

OUTLINE STUDIES IN BIOLOGY

Editor's Foreword

The student of biological science in his final years as an undergraduate and his first years as a graduate is expected to gain some familiarity with current research at the frontiers of his discipline. New research work is published in a perplexing diversity of publications and is inevitably concerned with the minutiae of the subject. The sheer number of research journals and papers also causes confusion and difficulties of assimilation. Review articles usually presuppose a background knowledge of the field and are inevitably rather restricted in scope. There is thus a need for short but authoritative introductions to those areas of modern biological research which are either not dealt with in standard introductory textbooks or are not dealt with in sufficient detail to enable the student to go on from them to read scholarly reviews with profit. This series of books is designed to satisfy this need. The authors have been asked to produce a brief outline of their subject assuming that their readers will have read and remembered much of a standard introductory textbook of biology. This outline then sets out to provide by building on this basis, the conceptual framework within which modern research work is progressing and aims to give the reader an indication of the problems, both conceptual and practical, which must be overcome if progress is to be maintained. We hope that students will go on to read the more detailed reviews and articles to which reference is made with a greater insight and understanding of how they fit into the overall scheme of modern research effort and may thus be helped to choose where to make their own contribution to this effort. These books are guidebooks, not textbooks. Modern research pays scant regard for the academic divisions into which biological teaching and introductory textbooks must, to a certain extent, be divided. We have thus concentrated in this series on providing guides to those areas which fall between, or which involve, several different academic disciplines. It is here that the gap between the textbook and the research paper is widest and where the need for guidance is greatest. In so doing we hope to have extended or supplemented but not supplanted main texts, and to have given students assistance in seeing how modern biological research is progressing, while at the same time providing a foundation for self help in the achievement of successful examination results.

J. M. Ashworth, Professor of Biology, University of Essex.

Human Genetics

J. H. Edwards

Professor of Human Genetics,
Infant Development Unit,
Birmingham Maternity Hospital
Birmingham

LONDON
CHAPMAN AND HALL

A Halsted Press Book
John Wiley & Sons, New York

First published in 1978
by Chapman and Hall Ltd
11 New Fetter Lane, London EC4P 4EE

Typeset by Preface Ltd., Salisbury, Wilts.
and printed in Great Britain at the
University Printing House, Cambridge

ISBN 0 412 13170 6

Distributed in the USA
by Halsted Press, a division
of John Wiley & Sons, Inc., New York

Library of Congress Cataloging in Publication Data

Edwards, John Hilton.
Human genetics.

(Outline studies in biology)
"A Halsted Press Book."
Includes bibliographical references and index.
1. Human genetics. I. Title.
QH431.E35 573.2'1 77-21582
ISBN 0-470-99096-1

Contents

Introduction

The awareness that children resemble their parents probably arose early: it was certainly obvious to the authors of Genesis and assumed to be self-evident to the audience of Homeric epics.

The mechanism by which these resemblances were manifested, and the extent to which inborn variation could be either amplified or suppressed by natural environments, or by such artifical environments as those imposed by education or medical interference, remained wholly unknown until the major components of these mechanisms had been observed or inferred.

In human genetics, as in astronomy, the natural curiosity of man about his past and future led to various extensive beliefs, varying from the flatness of the earth and a belief in horoscopes to the irrelevance of biological parentage to intelligence, behaviour or disease.

This book aims to provide a basic introduction to the mechanism of heredity, and to the ways in which these mechanisms may fail and lead to disease.

The emphasis is on man, but since the basic mechanisms of man differ little from those of other mammals, it is hoped that the book may be of some use to those whose primary interests are related to the mouse or the farmyard. It is based on a course of lectures given to the medical students in the University of Birmingham; this course included lectures on cytogenetics by Dr. Alan McDermott, who has written in this series.

In general the approach will be mechanistic and deductive, with the main emphasis on the necessary consequences of self-reproducing and self-copying genetic elements which, in 1978, we may regard as being 'known' to exist.

This is not the most usual way to develop this subject, since medical students traditionally start on the adult in the dissecting room and work from the obvious to the unseen. However, the time now seems ripe to follow geometry, astronomy, physics and chemistry, in defining a basic set of components, and building from these until, like a child with a set of Lego, recognisable structures with self-evident properties can be developed.

1 The units of inheritance

A simple model of inheritance is to suppose a large number of units, each determining some character, such as eye-colour, or some part of a character, such as the adding of a hundredth of a cubit to the stature, which merge at conception, and then split at reproduction. This model, which was implicit in much early work, adequately explained the similarity of parent and child.

This blending of the hereditary units did not explain the dissimilarity of brothers and sisters, or the persistence of hereditary variation. Mendel's model in which the units, or, factors as he called them, remained distinct between conception and reproduction, and were passed on intact and independently, each unit having a half-chance of entering each gamete, had the important distinction that only their effects were 'blended', and then only sometimes. This provided a model in which unchanging units could maintain a stable variation, and explained both the similarities and the dissimilarities of children, or, in Mendel's case, of the progeny of the sweet-pea.

Mendel also advanced the concept of unequal effect, so that where two different factors were inherited, the effects of one might so dominate the other that its partner would not be evident. He called these effects dominance and recessivity, and used the adjectives dominant and recessive for the factors.

He introduced the use of upper and lower case letters, the upper being dominant, and introduced the diagram (Fig. 1.1) by which 'factors' can produce 'effects'.

This diagram, in which two gametes lead to all possible pairs of alleles, can also be used backwards, so that all possible pairs of genes can lead to two types of gametes (a and A). The output gametes, unlike the input gametes, would of course both be male or both female.

Mendel summarized this in two 'laws'. The first stated that any gamete produced by an organism is equally likely to contain the contribution (factor) from either the male or female gamete that gave rise to that organism. That is, there is an equal splitting (spaltung) of the genetic material, a

		Female contribution	
		A	a
Male contribution	A	AA	Aa
	a	aA	aa

Fig. 1.1 Mendel's diagram

word which Bateson intentionally mistranslated as segregation, since he could not accept the premise of passivity. Mendel's second law stated that pairs of loci are also conveyed independently, a law which is only true of loci not mechanically associated by lying on the same chromosome, that is, of loci which are not linked. As all advanced organisms have many chromosomes, most pairs of loci are unlinked.

The two 'factors' A or a coexist after fertilization; when the effect of A is manifest independently of a, the organisms with AA, Aa and aA being indistinguishable when A is dominant to a.

This model was not applied widely until early this century, over forty years after publication in an obscure journal. The model was at first considered incapable of explaining graded or continuous variation, such as height; this inadequacy was removed by Fisher's paper of 1918 which showed that a model with many independent factors could explain with precision the inheritance of height in man.

The nature of the 'factors' was inferred to be that of producers of 'ferments' by Garrod, an English physician; the 'one gene one enzyme' model was later substantiated by direct experiments on *Neurospora*, a fungus. The development of microscopy, both optical and electronic, and of methods for the separation, detection, and characterization of both the genetic determinants, or factors, and their products, has now provided us with an elaborate hierarchy of genetic elements within which Mendel's factor, now termed a 'gene', may be regarded as the unit of protein synthesis. As each gene has a position or locus, and as variant forms (known as 'alleles') usually exist, it is convenient to avoid the word 'gene' wherever possible, since it is often used ambiguously for either locus or allele, and to speak of loci and alleles. Fig. 1.2 shows the hierarchy of genetic elements: the numbers should not be regarded as exact (in man there are 23 chromosomes per haploid set, 20 in the mouse, 7 in a muntjak and 80 in the rhinoceros – the least and the most in numerical terms). All mammals have, within the precision of the methods, the same amount of DNA, between 3 and 4×10^9 nucleotide pairs per gamete.

In addition there are genetic elements of unknown origin or function composed of distinct varieties of DNA, distributed irregularly among the majority type and representing some 3% of the total in man, and some 10% in the mouse. The finding of these fractions, which vary greatly between related species, is the major genetical discovery of the last decade.

These fractions are known as satellites, since they usually descend alongside the main type of DNA at ultra-centrifugation, due to having different proportions of the four nucleotides, which are all of different molecular weight. They will not be discussed further since they have not, as yet, been related to either normal or morbid variation; it is not even clear whether they are functional and necessary components, as opposed to ingenious parasites which have imposed themselves on the thread of life.

Self-reproducing elements of DNA are also found in such organelles as mitochondria; again we will ignore these in this outline since no consequence of their variation is known. Mitochondria appear to breed true, although many, if not most, of their associated enzymes are coded by nuclear DNA. Most, if not all, mitochondria are of maternal origin: it is not known if the few mitochondria at the root of the sperm tail survive conception.

1.1 The nucleotides

The elements of DNA, like the units of a zip-fastener, have a unit, the. base, which mates with its inverse, and a structural component which holds the base in line through the desoxyribose segment being coupled to its neighbour by a phosphate bond. A zip-fastener contains only two types of connection, a bump and a dent, which alternate, each unit having both a bump and a dent. If units were flat on one side, and had either a bump or a dent on the other, a simple structure would exist in which the strands could be separated and then each partner reconstructed from a set of spare units of each kind, leading to two identical 'zips'. The process would be repeated indefinitely and, for a zip of n elements, there would be as many as 2^n different variations.

The genetic code contains four distinct elements, adenine, thymine, guanine and cytosine, or A, T, G and C; since thymine and cytosine have pyrimidine bases and adenine and guanine purine bases, and the former are smaller, having one ring, while purines have two, we may write these as:–

A t G c

The unconventional use of small letters defines the purines. Compatible pairs of nucleotides are those in which one is a pyrimidine and one a purine. There are also differences in bonding, cytosine and guanine having three hydrogen bonds and adenine and thymine only two; we may therefore write the code as

A t **G c**   where heavy type defines a triple bond.

A stable pairing is only possible if the units have the same total length (that is, since pyrimidines and purines are of different sizes, if there is one of each) and if the hydrogen bonds are equal in number between each pair so that, given one strand of:–

A t **G c g** A t **c** t,

the other strand must be:–

t A **c G c** t A **g** A,

and each strand can therefore replicate, and, if the replica replicates, the original is reproduced. Replication implies a distinct inverse copy, as in printing or photography, and is not a synonym for reproduction. This is the genetic code. The information content for n units can be as high as that implied by 4^n different possibilities, or, in 'bits', where a 'bit' is the information implicit in a single choice between two possibilities, up to $2n$.

This structure, which may be visualized as a zip fastener, or as a ladder whose rungs are the bases and whose supports the backbone of phosphate linkages, is folded into a helix, the 'double helix'. However, at an elementary level, its properties are adequately covered by a linear zip or ladder model.

1.2 The gene

The nucleotide sequence is both informative, through its variation (the rungs of the ladder), and coherent, through its uniform phosphodiester linkages. In organisms with a naked chromosome, or chromoneme, the gene is defined syntactically by the information it contains, rather than being a distinct functional unit on a structural basis, as in the 'bead on a string' model of earlier days.

Even though there is no evidence of any structural discontinuities between genes, it is often convenient to retain the word locus for the site of this functional unit.

The discontinuity implicit in the gene is manifest at the functional level, since a continuous string of genic material, or DNA, will direct the synthesis of a series of discontinuous strings of amino-acids, the proteins.

We may define the units coding for proteins as genes. Proteins are usually 200–300 amino-acids long, or compounded of units of this length, and three nucleotides are needed to specify an amino-acid, some 600–900 nucleotides are needed for synthesis, in addition to others for starting, stopping and other poorly understood functions. The haemoglobin coding segment for rabbit haemoglobins is known to be at least 200 nucleotides longer than the 450 necessary for the synthesis of 150 amino-acids.

Since each gamete conveys over 3×10^9 nucleotide pairs, and most proteins require less than 10^3 for their coding, there is room for 3×10^6 or three million genes. Even if generous allowance is made for 'structural' elements allowing for chromosomal pairing, for inert spacers, for repeated units synthesizing ribosomal RNA, and for satellites, there is adequate room for a million genes – or for 20 000 on the smallest human chromosome.

This may seem a somewhat generous supply of proteins but is by no means impossible; most of these would be involved in subordinate or minor activities, and many, like words in a large dictionary, would be of purely historical interest. Furthermore, many genes may be duplicated, or coded at multiple sites, as is known to occur in human haemoglobin.

1.3 Structural units

In bacteria many examples are known in which the enzymes involved in serial metabolic actions are coded at a series of neighbouring loci, and the whole genic sequence of protein-forming genes, sometimes called structural genes, is under the integrated control of a connected series of 'regulator genes'. Only a minority of genic loci are integrated in this way.

Few examples of integration between the location of genes defining associated functions have been found in mammals, and many counter–examples are known. The chains of the main haemoglobin in both fetus and adult are coded from distant loci: in mice they are known to be on distinct chromosomes. The heavy and light immunoglobulin chains are similarly distinct.

On the main glucose pathway phosphogluco-dehydrogenase is coded on chromosome 1, glucose-6-dehydrogenase on the X-chromosome.

Doubtless some coordinated and serial systems have retained the relationship of their remote past, but few are known. In the absence of knowledge of 'regulator' genes the term 'structural gene' is unnecessary. Unless otherwise specified, gene, or allele, will imply a unit coding for a protein.

1.4 Repeated units

Substantial numbers of ribosomes, the standardized organelle on which proteins are coded, are needed in all cells, and are replicated directly from the DNA from numerous sites, at each of which many identical segments of DNA are present. Transfer RNAs are similarly coded from repeated units. There is also evidence of other repeated units, including ones too short to be able to code a protein. These short units are particularly mysterious, since they are both dispersed throughout the genetic apparatus, or genome, and highly specific by species.

1.5 Chromosome arms

Most human chromosomes have two arms, and are usually drawn with the shorter above the larger. In mice, cows, sheep and horses almost all the chromosomes have one arm far longer than the other, which may not even be visible. The reason for this is unclear. The reason for the chromosomes in any species to resemble each other, or, as in man, to form distinct groups, is also unclear.

1.6 The haploid set of chromosomes

This is the unit of sexual reproduction.

1.7 The diploid set

This, the nucleus in normal mammals, is the unit of cellular reproduction.

1.8 The genetic hierarchy

This set of mechanisms forms a hierarchy, each element being composed of several or many elements from lower in the hierarchy.

All these elements must reproduce and, in doing so, all are liable to error or mutation, some of which will persist. In the absence of an error, evolution would not occur. In the presence of too much error the species afflicted would not be competitive with its rivals. All these elements mutate at different rates, the larger elements in the hierarchy mutating faster. At no level of the hierarchy is the mutation rate likely to average more than once per unit per generation.

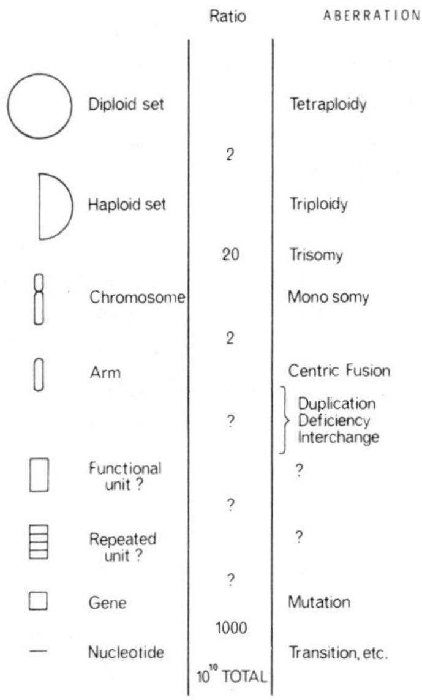

Fig. 1.2 The genetic hierarchy.

1.9 Protein synthesis

The primary functions of the genetic apparatus are, firstly, to duplicate itself and, secondly, to direct the synthesis of impermanent structures, the proteins, which will permit its survival. Protein synthesis is effected by the opening up of a segment of DNA to lay bare the bases; the synthesis of a matching chain of RNA from four components (adenine, guanine, cytosine and uridine: thymidine is not found in RNA); the release of this chain from the DNA; its transport from the nucleus to the cytoplasm; its capture by the two sub-units of a ribosome, which come together, trapping the chain by its beginning; and then the steady threading of this chain along the groove between the ribosomal sub-units, during which transport RNA (small molecules with a recognition site for RNA triplets at one end and a loosely attached amino-acid at the other), effects the translation of the genetic code to produce the linear segment of amino-acid, which is of course protein.

The processes are well understood in bacteria where transcription and translation take place alongside the chromosome. In nuclear organisms the transcription appears to involve a very long chain with regions which are added but not transcribed, this long chain then being segmented into transcription units which are then exported from the nucleus to the cytoplasm: other parts are apparently degraded in the nucleus.

In some unusual cells, such as immature mammalian red cells, messenger RNA may last many days, leading to the synthesis of millions of molecules at a rate of many nucleotides a second, or several protein molecules a minute per ribosome.

1.10 Disorders of transcription

Any defect in the enzymes involved might be expected to be incompatible with life. Since no cell transcribes more than a small minority of its DNA, differentiation must involve defining the segments to be transcribed; deficient transcription is a plausible cause of conditions in which the protein formed is normal in quality but deficient in quantity. Some beta thalassaemias, a class of conditions due to deficient synthesis of haemoglobin, are due to the degradation of the messenger before it has had time to be translated an adequate number of times.

1.11 Translation errors

Translation errors will arise when any mutant transfer RNA provides an unusual translation. In higher animals such a mutation is unlikely to be sufficiently viable to become established, and is likely to be present as a so-called lethal dominant. Certain nucleotide changes might lead to a messenger RNA which was so twisted upon itself that it could not be unravelled. This is a theoretical possibility which, while likely to arise, would be very difficult to detect.

2 The chromosomes

Man has 46 chromosomes, of which 44 are evenly paired in both sexes, and two sex chromosomes which are different in the male.

The chromosomes which are not sex chromosomes are called *autosomes*. The sex chromosomes are sometimes called *allosomes*, but this is not in common use in man, and the terms X-chromosome and sex-chromosomes are often used ambiguously and even erroneously. The chromosomes are usually observed by culturing lymphocytes in the presence of a stimulant and then allowing the dividing cells to accumulate in cell division, or *mitosis*, by adding a substance which blocks its completion. The chromosomes which are about to separate to form the nuclei of the daughter cells are then seen as double structures connected at their centromeres, the centromere being the constriction visible on all chromosomes and connected to the fibres by which the chromosomes are caused to separate after division. The metabolically active chromosome is a very attenuated single strand which forms almost the entire stained matter in the cell nucleus when observed on conventional microscopy.

On displaying the chromosomes by photography, followed by cutting out each chromosome and ranking them by size, shape, and staining characteristics, the characteristic karyotype is obtained (Fig. 2.1). Using modern staining techniques each chromosome can be identified, and, due to small variations in many chromosomes, any individual can usually be distinguished from any other individual.

The chromosomes may deviate from the normal pattern either by being too few or too many, the *numerical abnormalities*, or by individual chromosomes being abnormal, the *structural abnormalities*: in some cases, as when two chromosomes are fused together, structural abnormalities may lead to a numerical abnormality.

It is convenient to discuss the abnormalities of the autosomes and sex chromosomes separately.

2.1 The autosomal aberrations

About 20% of conceptions are known to die *in utero* and to be aborted in consequence; there are usually some surviving cells and if these are established in tissue culture the chromosomes can be displayed. It is now clear that most of these abortions have numerical chromosomal errors, the commonest being an additional autosome, a missing sex chromosome, and the addition of a complete set of haploid chromosomes, so that 69 chromosomes are present, a condition called *triploidy*.

If the newborn are studied, either by survey or by clinical observation followed by the chromosomal study of unusual children, it is clear that only a minority of conceptions with abnormal numbers of chromosomes survive, the mortality varying from over 99.9% in triploidy and most varieties of extra autosome (trisomy) to about 98% in fetuses with an X-chromosome unaccompanied by another sex chromosome, to a mere slight majority in those with an extra chromosome 21, the survivors of which will have *mongolism*, often called *Down's* syndrome after the English physician who described it in 1866. The pattern of this mortality is shown in Fig. 2.2. There is a remarkable discrepancy between the mortality of conceptions and the morbidity of the survivors. Most children with a single X-chromosome slip through even expert neonatal scrutiny, and, although dwarfed and unable to mature sexually, are usually of average intellect. In mongolism almost all are identified at birth, and intellectual development is normally bounded by the limits of being able to dress but not to read.

Rather more than one baby in a thousand has trisomy 21 or mongolism. The only other trisomic conditions with a frequency exceeding once a paediatrician-lifetime are trisomy 18 and trisomy 13, with frequencies of about one in three thousand and one in five thousand respectively. Both are more severe than mongolism, as is to be expected as the offending chromosome is also bigger, and in the absence of enthusiastic resuscitation followed by tube feeding, natural death usually occurs in days or weeks.

Mongolism, or trisomy 21, provides the bulk of the disability due to chromosomal disorders, and, as will be discussed later, may arise in various ways. However, about 96% of cases are due to a simple trisomy in which the individual chromosomes are normal, and the incidence of this disorder in the newborn is greatly influenced by maternal age, rising from

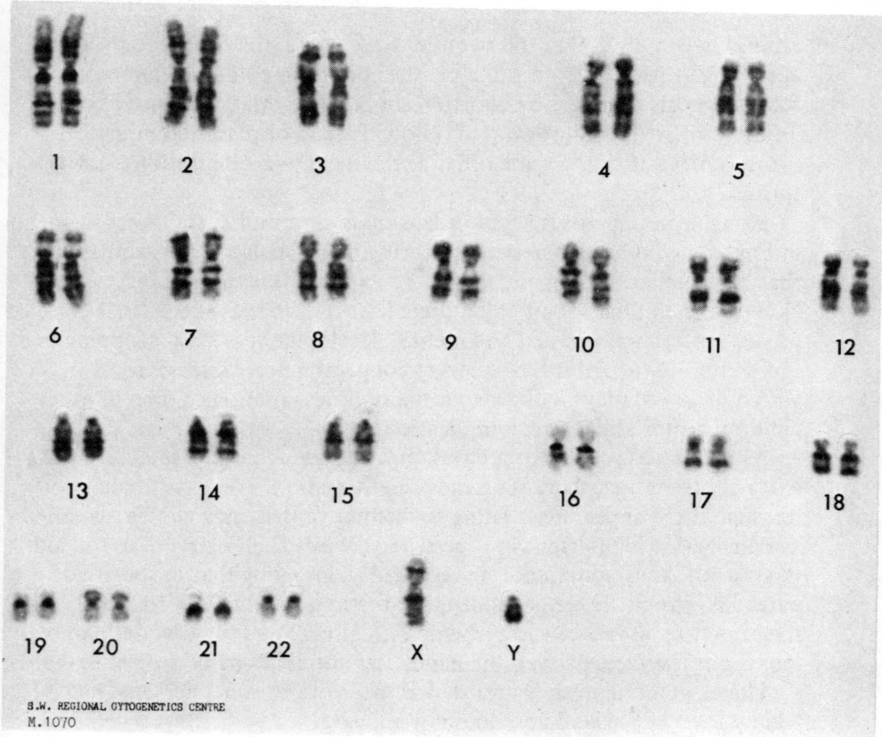

Fig. 2.1 Human male karyotype.

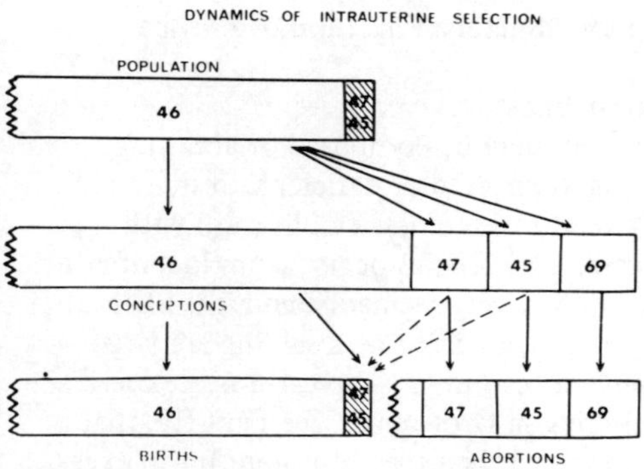

Fig. 2.2 Dynamics of intrauterine selection.

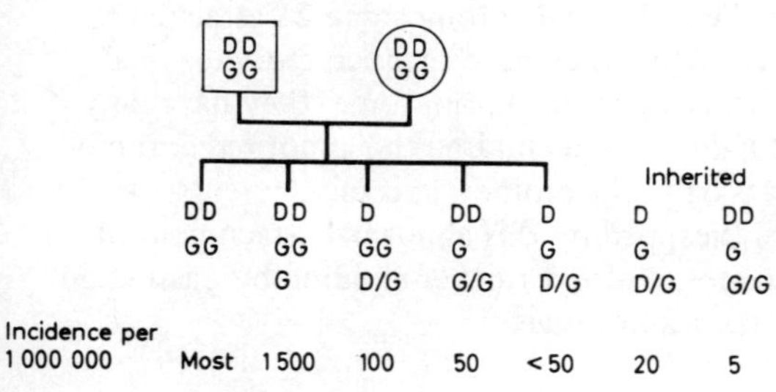

Fig. 2.3 Approximate incidence of types of mongolism (Down's syndrome) per million births.

around one in 10 000 in the twenties to one in a hundred at forty and about 3% at forty five. In principle the condition could be eliminated by fetal diagnosis followed by abortion: in practice this procedure has the limitation that the abortion rate following fetal or placental injury at amniocentesis is of the same order as the detection rate until the late thirties.

Mongolism imposes severe burdens on most parents, and most women in England who have experience of bringing up such a child seem anxious that any further pregnancies should be exposed to amniocentesis with a view to the elimination of any affected fetus. So far as the child is concerned suffering is unusual and mental development is slow, stopping in a few years. About a third have severe congenital heart disease, most of whom die in infancy: leukaemia is many times commoner than in other children, but is still a rare complication.

As is to be expected from development in which every cell carries an extra chromosome, there is a general deformation, every part being unusual, some appearances being so distinctive that they can be classified as signs with a high diagnostic specificity. While such signs are useful aids to students, they should not be regarded as implying that unspecified parts are normal. These generalized distortions are difficult to specify, in the same way as those features which usually allow sex to be defined by looking at the face, or even the hands, cannot easily be conveyed in words.

The term mongolism is regarded as offensive to some persons, but it is widely used in England and does not appear to cause distress to parents. In the United States, which has a substantial Chinese population, including several distinguished geneticists, the term has been almost completely replaced by Down's syndrome. Down considered the disorder to be a residual genetic trait following the Mongolian invasion of Europe.

2.2 Structural autosomal abnormalities

The karyotype, or chromosome set, may be composed of abnormal chromosomes due to a part being duplicated or deficient, or due to an interchange of parts between two chromosomes. Duplication without some associated deficiency is rare. Deficiencies occur, from loss of either the ends of chromosomes or of intervening segments, and it is likely that only a minority of deficiencies large enough to be disabling are large enough to be detectable on conventional microscopy. Human experience in man is consistent with the finding in Drosophila, the fruit fly, that a deficiency is likely to be more harmful than the equivalent bit in excess. About 4% of mongols have the additional chromosome 21 attached to another chromosome, so that, while they have as much chromosomal imbalance as if they had a freelying extra chromosome, they have only 46 chromosomes. In about half of these children the abnormal chromosome comes from a parent, usually the mother, and such parents, although not themselves handicapped by this abnormal attachment of normal chromosomes, are likely to afflict further children by passing on more than an exact half of their autosomes.

Substantial rearrangements are found in well over one in a thousand persons, and some of these lead to a degree of imbalance in their children which, while not sufficient to lead to fetal death, is sufficient to disturb development. Further details of these rarer and more complicated abnormalities may be found in 'Cytogenetics' in this series.

The only practical point in the detection of such defects is the appreciation that a generalized disturbance sufficient to depress intellectual development is likely both to depress and to distort physical development, and since in such cases a normal parent may have two abnormal chromosomes and the abnormal child only one, the parental chromosomes should always be analysed alongside the child's.

2.3 Normal variations of the sex chromosomes

Sexual differentiation in man is determined by the sex chromosomes, the presence of a Y-chromosome influencing the gonad, as a result of which it both encourages and depresses the development of various embryonic

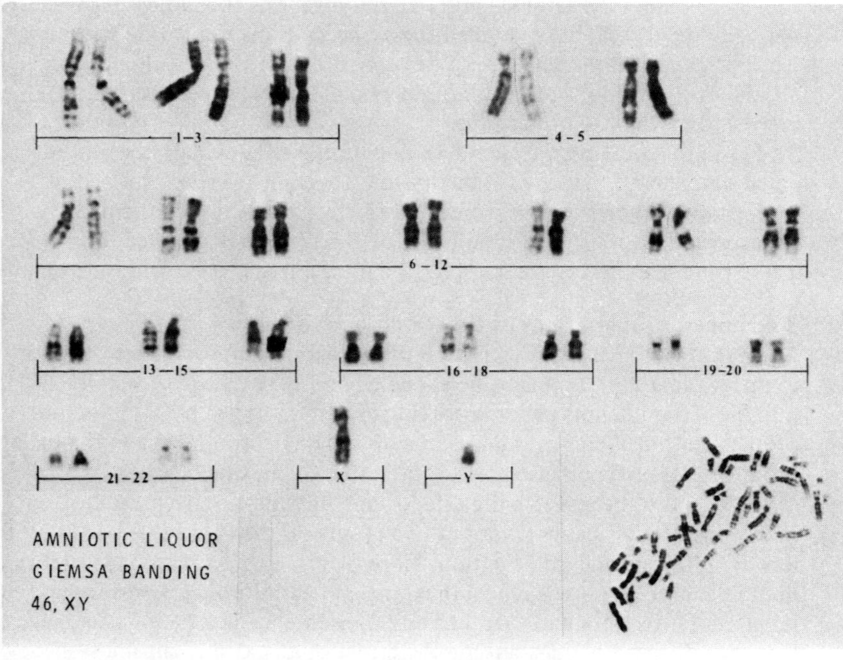

Fig. 2.4 Normal acrocentric chromosomes and the common varieties of translocation mongolism. (Carrier detection – chromosomal.)

tubes in its immediate neighbourhood, and, through the secretion of testosterone and other hormones, induces profound and irreversible effects on the various organs, including the less intellectual parts of the brain, whose development ensures the characteristic physical and psychological sexual differences.

Since the X- and Y-chromosomes differ greatly in size the mechanisms involved in their distribution in the male gametes, or sperm, are likely to be complicated, and, in consequence, likely to go wrong. Since the greater part of the physical and mental equipment of both sexes is virtually identical, and since the X- and Y-chromosomes differ so substantially, it is reasonable to suppose that there should be some mechanism through which the two X-chromosomes in the female should have their activities reduced to a level comparable to that achieved by the solitary X-chromosome in the male. In 1958 Bertram and Barr discovered a small body in the nucleus of the female cat, and later found a similar body in various cells in human females, especially scrapings from the mouth (the buccal smear). The reasonable assumption that this was one of the two X-chromosomes in an inactive state was confirmed by findings in girls with only one X, in women with three X-chromosomes, and in men with two or more X-chromosomes.

The further assumption, advanced by Dr. Mary Lyon (and therefore known as the Lyon hypothesis) that the active and inactive X-chromosomes are randomly defined early in embryogenesis and, when once defined, breed true, has extended these observations to provide a framework of expectations which has been consistent with most, though not all, the manifestations of both normal and abnormal genetic variation conveyed on the X-chromosome.

The cytological basis of sex determination would explain the almost equal numbers of the sexes at birth if the abortions were also equal by sex; detailed studies using chromosome analysis have not confirmed the widespread but poorly justified belief of a male excess at conception. There is no need to invoke complicated mechanisms to explain this equality.

2.4 Numerical aberrations of the sex chromosomes

Numerical errors in the distribution of the sex chromosomes are common in man (Table 2.1). As might be expected from the nature of sex determination, the phenotypic or apparent sex is determined by the presence of one or more Y-chromosomes; an additional X-chromosome in the male produces a specific disturbance to the testis which is inconsistent with fertility, and some less specific effects, one of which is a slight reduction in intelligence, probably by about five points on the IQ scale, or rather less than half a standard deviation. More than one extra X has a severely disabling effect on intelligence. In women an additional X seems to have no effect on fertility and even less effect on intelligence than in the male. It is very difficult to substantiate the effect of an extra X-chromosome, which is common at birth but not found with a grossly disproportionate frequency in girls with disturbances of growth, puberty, intellect or behaviour.

Table 2.1 Common sex chromosome abnormalities.

XXY	1/500	males	Sterility, slight reduction in IQ.
XXX	1/500	females	Slight reduction in IQ.
XYY	1/500	males	Tallness, slight increase in aggressiveness.
XO	1/4000	females	Sterility, shortness.
	1/100	abortuses	

Approximate incidence of numerical sex chromosome abnormalities at birth and at conception. (Only the XO is substantially rarer at birth, 98% aborting.)

An additional Y-chromosome has, as might be expected, the effect of accentuating the manifestations of maleness, and especially the differences in height and aggression. The normal male is, on average, 12 cm or 1.7 standard deviations taller than the female, and is responsible for almost all criminal activity. The majority of XYY males conform to the limits of aggression tolerated by society, but they have an undoubted, and hardly unexpected, over-representation in those convicted of aggressive behaviour and in those who are tall.

Individuals born with a solitary sex chromosome, and this is always an X-chromosome, are usually fairly normal in infancy and childhood but grow slowly in later childhood and do not achieve any pubertal change unaided. A minority have various well marked features, including the rather froglike neck, originally described by Turner. They are infertile.

Various mosaic forms may arise, and some, such as XO/XY mosaics, in which some but not all cells lack a Y-chromosome, may have the misfortune to occupy an intermediate position between the usual male and female forms.

These various deviations are of practical interest since they present problems if discovered during fetal or neonatal life, and may lead to responses varying from fetal death to unnecessary psychiatric supervision and parental anxiety. So far as an extra X in a woman, or an extra Y in a man, is concerned these can almost be regarded as within the broad limits of normal variation.

3 The nature of the genes

The only adequately understood units of inheritance are the chromosomes and the genes. While the chromosomes are the genes, in the sense that they are structures whose active components are those parts of themselves which can both copy themselves and also release transient copies through which they direct and dominate cellular activity, the relationship of these parts to the whole is quite obscure. The chromosome is a cellular *organ* which has a certain degree of autonomy, approximately half its mass consisting in histones and larger proteins, including actin and myosin, and these substances are intimately entwined with the DNA spine in such a way that, while chromosomes from all cells look identical, they clearly differ in that the obvious differences between different cells must be the consequence of differential inactivation of parts of their constituent chromosomes.

These are matters of grave difficulty which will not be pursued further. The problem of why brothers or sisters differ from each other notwithstanding a common parentage and environment is substantially solved: the problem why distinct cells either become distinct or remain distinct is not. This is the problem of *differentiation* whose solution is unlikely to precede an understanding of how the constituent parts of the chromosome are assembled together.

We will ignore these difficult matters and make the traditional assumption that the chromosomes are strings on which the genes lie, like beads, in a defined order. We will ignore the problems of differentiation for, so far as many of the practical issues of human genetics are concerned, the implications are not greatly changed by the fact that most genetic activities are unevenly expressed in different tissues: indeed, it is this unevenness of expression which gives rise to tissues.

Obviously mechanisms of such complexity will have many ways of going wrong, and defective or premature differentiation is likely to be responsible for a substantial proportion of disease. In this short book we will merely consider the classical model of the gene, for this is not only sufficient to provide a basis for almost all the simple and discontinuous variation observed with, for example, blood groups, but also provides a sufficient explanation for both genic disorders, such as cystic fibrosis, and genetic variation in such measures as height, colour, intellect, or predisposition to disease.

The gene has the capacity to reproduce itself at cell division and to influence the cell within whose nucleus it lies by disseminating transient copies of itself. That is, it can take part in *replication*, by which an inverse copy of itself is made, and from which, by further replication, it can be reconstituted, or it can take part in *transcription* by which an inverse copy on another molecule (RNA), known as messenger RNA, is then used for the *translation* of its message to produce a protein. This basic mechanism, which is common to all DNA-containing organisms, and which was originally studied in bacteria, is considered in detail in *Protein Biosynthesis* by Alan Smith. We will content ourselves with a simple diagram which is intended to emphasize the differences between replication and transcription (Fig. 3.1).

The protein molecule is the primary genetic product and the mass of proteins produced, all of them the consequence of transcription and translation, are the immediate cause of our individuality, and provide a sufficient basis for the distinctions between both individuals and species.

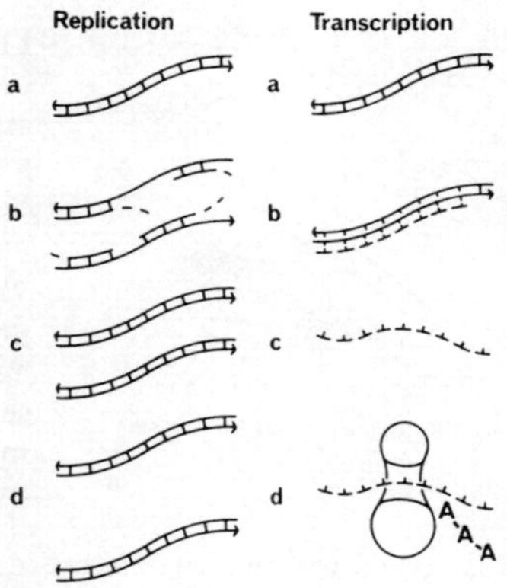

Fig. 3.1 Replication and transcription.

Replication	Transcription
a. Resting duplex chromosomal segment.	a. Resting duplex chromosomal segment.
b. Chains separate and DNA units attach to exposed parts on each strand and become attached to phosphate backbone to form new duplexes.	b. Segment separates; RNA units attach to exposed parts and join up one strand.
c. Duplex chains separate.	c. RNA strand separates carrying 'imprint'.
d. Chains form new daughter chrom-osomes each of which is 'half-new'.	d. Strand reaches cytoplasm and codes protein (A-A-A) on ribosome.

Events a, b, and c relate to replication and transcription; d refers, on the left to separation at cell division, and on the right to translation.

Most proteins are enzymes; a few are structural, although most structural elements, such as bone, are largely the products of enzymes. Others are conveyors: examples of these include haemoglobin, haptoglobin and transferrin.

The general nature of the way in which DNA, with its constituent genes, each with a half-life of millions of years between any change in structure, defines and determines the proteins whose half-lives are usually measured in days, is shown in Fig. 3.1. We may attempt to look at our biochemical individuality in another way by means of a diagram which represents metabolites, whose interchange is catalysed by enzymes. Each enzyme pathway is paired, one track being catalysed by the enzyme coded by the paternal DNA, the other by the maternal. We ignore, for simplicity, the occasional problems of the male with his solitary X-chromosome.

The differences between individuals may be thought of as largely the consequence of differing amounts of constituents (represented by the size of the circles in Fig. 3.2) and these differences are largely the consequence of the variation in the activities of the parental enzyme. Obviously the environment is also in a dominating position, in that extremes, such as starvation or drought, can impose patterns of metabolism which are

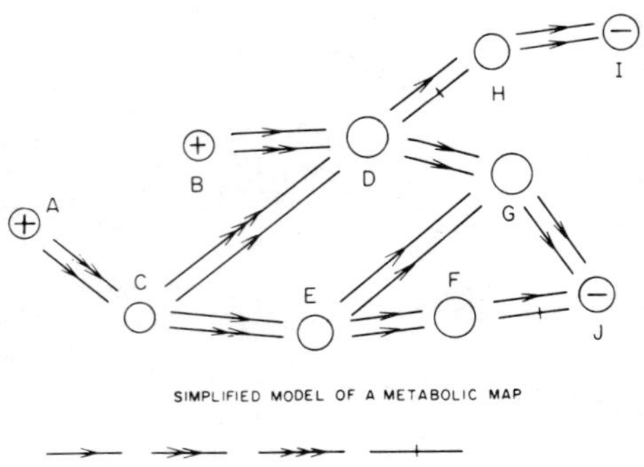

Fig. 3.2 Simplified model of a metabolic map. The arrows show the catalytic pathways between metabolites (a circle). These pathways are catalysed by enzymes coded independently by the paternal and maternal DNA. A serious genetic situation may arise if there is a pair of defective enzymes.

more extreme than those usual in the most extraordinary individuals in more equable conditions. However, for simplicity, we will consider variations in a fairly uniform environment.

The variations are largely the consequence of differing enzymatic strengths, and this poses the problem as to how these strengths are determined. The simplest mechanism, in which various numbers of genes are present, does not seem a common mechanism for determining enzymatic activity: in general only one locus seems to be present on each chromosome for each protein. Exceptions certainly occur, as in the human alpha and gamma haemoglobin chains. We could also have activity modulated by speed of transcription or translation, or by the stability of the messenger RNA, which is normally transcribed numerous times before its destruction. None of these mechanisms seem common as sources of normal variation, although the variation underlying some thalassaemias is related to the fragility of the messenger. The bulk of the variation within species, and, by implication, between species, seems to be due to qualitative variations in the DNA, as a result of which slight qualitative differences in the protein produced, usually a single amino-acid, affect its behaviour on such coarse procedures as electrophoresis, and can hardly fail to affect the finesse with which such molecules handle the intimate problems of catalysis, or the varying favours they bestow on differing substrates.

Any variation in enzymatic strength will vary the amounts of metabolites, in the same way as varying the flow of rivers connecting lakes will affect their levels. Furthermore, any effect will be both immediate and remote, the localization depending on the size and nature of alternative channels.

Where there is only a single metabolic route and a simultaneous block of both lines, one controlled by a paternal and the other by a maternal enzyme, this will lead to a gross excess on one side and a gross deficiency on the other. This is basically how most recessive disorders produce their effects. Either the individual is encumbered or intoxicated by some excess, as in Tay Sachs disease or phenylketonuria, or there is a lack of some necessary substance, as in albinism or the inability to produce adult haemoglobin.

In view of the complexity of metabolic maps only a minority of such blocks would be expected to lead to any severe and readily localized consequence.

Consequences will be severe where the pathways are devoid of alternative routes, and this appears to be common when speed is essential, as in such basic mechanisms as oxygenation, or the major energy pathways. As in the case of the major highways in road transport, the benefits of speed are associated with a complete dependence on a few routes. It is enzymatic failure on such trunk routes which provide the conspicuous genic disorders of man. Most are autosomal recessives, and manifest only if both gametes convey a defect.

A small proportion of proteins, and this proportion is about equal to the proportion of the DNA on the X-chromosome, or about 1: 20, are defined at X-linked loci, so that in the male there is only a single pathway.

The inborn variation between individuals would seem almost entirely due to variations in enzymatic strength due to variations in the alleles; it is on this variation that selection has acted in the past, and which distinguishes brother from brother and man from ape. This classical view, most clearly elaborated by Sewall Wright, has recently been challenged by Ohno, who considers the quantitative mechanisms of control of protein synthesis more important than the qualitative variation in what is controlled.

4 Mendelian variation

The origin of inborn variation resides largely in variation at the loci, so that many pairs of loci contain different genes, or alleles; and the origin of these differences lies in mutation. Such mutations occur individually at a low rate, perhaps of the order of 10^{-5} to 10^{-6} per generation. That is, the average interval between mutations at any locus is some millions of years, so that each gene may be regarded as having a half-life of many millions of years, or of the same order as the half-life of many mammalian species, possibly including our own.

This variability, without which evolution could not occur, since selection would have nothing to select, arises blindly, and most such mutants would be expected to be disadvantageous in the same way as most misprints would degrade an article, a poem, or a musical score. But not all misprints would be disadvantageous.

It is against this background of mutation and selection as continuous processes that both normal and abnormal genetic variation must be interpreted.

Pathological variation arises both because the mutational mechanism is blind, so that most variants are relatively defective, and have to be weeded out by selection, and because, with so much variation, individually minor components of variation may, by chance, sometimes be so consistent in the direction of the variation they impose that the organism may be at the limits of the tolerance of the environment, or of the tolerance of other members of the species.

A disorder is said to be Mendelian if it is the consequence of a factor which obeys Mendel's first law. In practice it is unusual to include chromosomal abnormalities within this definition, which is normally restricted to disorders whose familial pattern is consistent with their being determined by some factor which is not related to any visible variation on optical microscopy.

We therefore exclude, by convention, the chromosomal disorders and restrict ourselves to the remainder, which we may term genic. In some of these, as achondroplasia, or Duchenne's muscular dystrophy, there are reasons to suspect that the factor which is abnormal, or deficient, is substantially larger than a gene, but such disorders can meanwhile be included within this term. Genic disorders, like normal genic variants, may be divided into dominants, autosomal recessives, usually termed recessives, and X-linked recessives. In man there is usually no evidence as to whether dominants conform to the classical usage of Mendel in which the heterozygote was indistinguishable from one homozygote, and the word is used as an abbreviation for 'manifest in the heterozygote'.

The problem of nomenclature have been simplified by McKusick's encyclopaedic work on *The Mendelian Disorders in Man*, in which all established Mendelian disorders, and some others, may be found.

The relationship of the genetic factor to disease must be necessary, although it may not be sufficient. Genic disorders, like other disorders, take time to develop: while many recessively determined only become evident a few days after the child is separated by birth from its mother, or only arise due to the failure of a normal adult protein to be formed, as in most haemoglobinopathies, others may depend on a specific environment. For example, those deficient in pigmentation may be unable to tolerate strong sunlight, those lacking trypsin antagonists in their blood will find smoking unusually hazardous, and those lacking the enzymes needed to degrade a common relaxant used in anaesthesia may be unable to breathe naturally for some hours after exposure. Some genetic disorders of gradual onset, most of which are dominants, may not become manifest until old age, and will obviously not be expressed in those who die from other causes first. However, such diseases are termed Mendelian if their manifestation in families is that expected if the appropriate environment is encountered and a sufficient age is reached.

4.1 Dominant disorders

Factors which are necessarily present are the cause of dominant inheritance, when a single factor must be derived from a parent, or consequent on mutation, and passed on with a 50 per cent chance to each child.

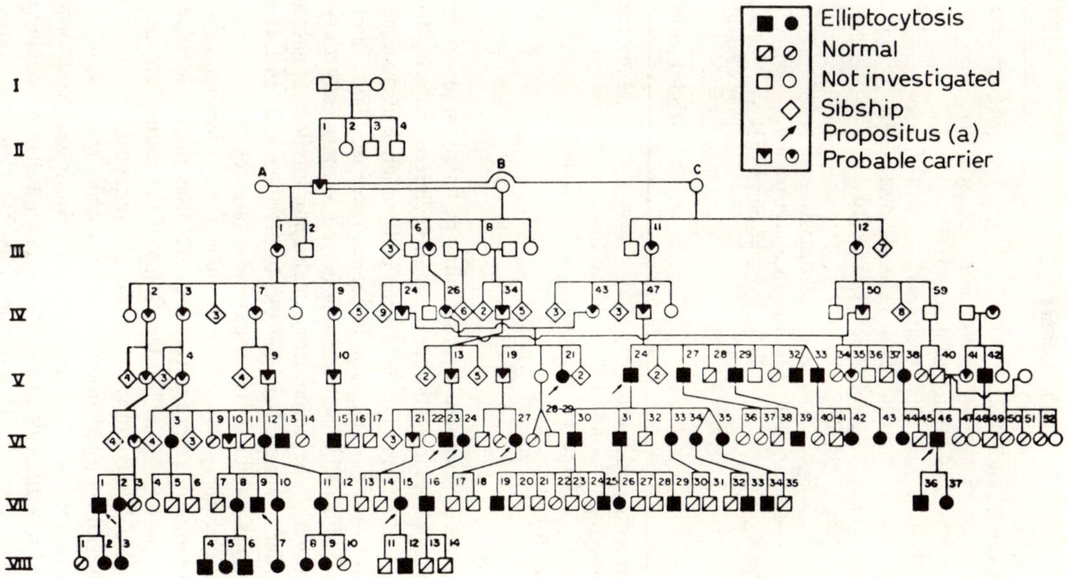

Fig. 4.1 Elliptocytosis in Iceland. (By kind permission of Dr. Olafur Jensson.)

Table 4.1 Approximate incidence, per million births, of genic disorders in man in Britain

Recessive		*X-linked recessive*	
Severe mental defect*	800	Haemophilia	100
Deafness*	500	Duchenne	200
Blindness*	200	Others	200
Fibrocystic disease	500		
Adrenogenital syndrome	100		About 500
Tay Sachs (1)	10	*Dominants*	
Sickling (2)	10	Blindness*	50
Thalassaemia (3)	20	Deafness*	50
Albinism	100	Huntington's chorea (4)	100
Phenylketonuria (4)	200	Neurofibromatosis	50
Others	200	Epiloia	50
		Achondroplasia	50
	About 2500	Dystrophia myotonica	50
		Polycystic kidneys (5)	100
		Others	100
			About 600

(1) Common in Askenazi Jews *Many forms
(2) Common in Africa; found in Greece and India
(3) Common in Italy, Greece, Africa and most of Asia
(4) Rare except in Europeans
(5) Asymptomatic cases far commoner

Evidence of the nature of these disorders may be based on one or more large pedigrees, or on a large number of small pedigrees. Elliptocytosis is an example of a fairly mild defect of red cells which leads to anaemia due to excessive wear and tear, and in consequence very large families have been found (Fig. 4.1). Achondroplasia, the commonest variety of severe dwarfism with otherwise normal health and intellect, is another dominant disorder in which social barriers to marriage and, until recently, physical barriers to childbirth, make extensive families rare.

Notwithstanding many statements to the contrary, dominant disorders, though numerous, are individually rarer than many recessive disorders and impose a smaller burden. Some of the commoner varieties are shown in Table 4.1. Polydactyly, although frequently quoted, following an early description by Maupertius, is not usually manifest in a parent. The nature of the offending factor in most dominant disorders is unknown, and it is difficult to see how such a mechanism could work. In achondroplasia, for example, a factor which has a mutation rate greatly exceeding that of any defined unit at the genic level seems able to disturb bone development but little else.

In the disorders in which there is a dominant predisposition, as those in which some factor associated with cell-membrane receptors is known to predispose strongly to damage from plant proteins, as in coeliac disease, or to an unusual infective disorder, or possibly an unusual response, as in ankylosing spondylitis or Reiter's disease, there is no problem in postulating a necessary factor. However, in such disorders only a minority of those predisposed are affected, and as there is good reason to regard the predisposing determinant as related to normal, rather than abnormal, variability, the term dominant disease should not be used in this context.

Dominance is, of course, a property of the technique used, and at the nuclear level all variation would be expressed if techniques allowed. In sickle cell disease, for example, the normal allele is dominant at the clinical level, but co-dominant at the laboratory level.

4.2 Recessive disorders

Other disorders may be consequent on the *absence* of some normal factor and, if conveyed on the autosomes, this absence will only occur if both parents fail to convey a normal factor. Such disorders are termed autosomal recessives, or recessives, since 'autosomal' is implied by default. In X-linked recessives the term is always written in full.

4.3 Normal mendelian variants

While normal and abnormal variation cannot be distinguished without reference to the environment, it is convenient to regard those variants which are not unconditionally harmful or undesirable as 'normal'. Many 'normal' variants including the ABO and rhesus blood groups, have only been recognized due to spontaneous incompatibilities, or due to those induced by pregnancy, by transfusion or by transplantation.

The ABO blood groups impose substantial differences in risk on recipients of mislabelled blood, as well as differences in the risk of acquiring cancers of the stomach and ulcers of the stomach and duodenum. Persons of group AB are fairly immune to transfusion errors;

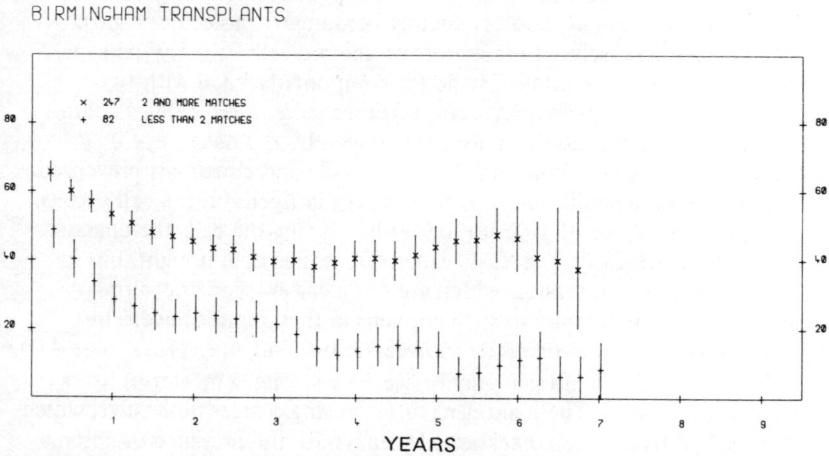

Fig. 4.2 Survival curve of kidneys after transplantation, with various degrees of incompatibility. Birmingham data. (By kind permission of Mr. A. D. Barnes and Mrs. Pauline Macintosh.)

those of Group O are most at risk, and also most susceptible to ulcers, but have a reduced risk to carcinoma of the stomach and salivary glands. There is no 'worst group' or 'morbid allele' in the way that the unit determining Huntington's chorea, or haemophilia, may be said to be unconditionally 'abnormal'.

Similarly, in rhesus disease, it is advantageous for men and babies to be rhesus negative and for women to be rhesus positive, and no rhesus allele can be said to be harmful in a meaningful way, although until recently it was a substantial cause of fetal and neonatal death.

The HLA (human leucocyte antigen) system provides a greater range of variation than any other system known in man, and some alleles clearly predispose to disease to such an extent that some individuals may be shown to be many times more likely than others to show some disorders. They include coeliac disease, in which the lining of the gut has its function disrupted by a protein present in wheat and rye, and ankylosing spondylitis, in which the joints, especially those of the spine, seize up or 'ankylose'. It also provides a rational basis for the matching of donors of solid organs, especially kidneys, although the advantages of matching, though clear, are relatively small (Fig. 4.2).

The wider implications of these normal mendelian variants, sometimes called 'markers', will be considered later (see Chapter 6) under 'Disorders of Defence'; meanwhile, since a basic knowledge of the ABO and rhesus system is obligatory for doctors, since ignorance could lead to fatalities, these will be described. The HLA system, although less important in practical medicine, will also be described briefly.

4.4 The ABO system

If blood from different persons is mixed, it will sometimes 'curdle'. This is due to the fluid part of blood from most people containing immunoglobulin molecules (antibodies), mainly pentameric molecules (IgM), which attach themselves to receptors on the red cells of many persons. Since the molecules contain five active components, each with two attachment regions, they may bond together pairs of red cells, a condition termed 'agglutination'. This is easily seen by the naked eye if bloods are mixed on a white tile or a glass slide and clotting is prevented.

The system depends upon the presence of antigens on the cell surface which may be 'A' or 'B' or both, or neither, giving the cells the characteristic A, B, AB or O. The 'O' is called 'O', rather than nought, but in fact means null. For reasons which are not clear everyone develops antibodies during infancy to such antigens as they do not have. Thus someone who has A antigens develops an anti-B and vice versa (Table 4.2).

In addition the A antigen may be weak or strong, a matter of little clinical importance. These antigens differ in single sugar molecules, which are attached to a protein backbone by enzymes, the presence or absence of these enzymes defining the antigen, and the two enzymes being coded for at the same locus, now known to be on chromosome 9.

Table 4.2

Inborn antigen	*Acquired antibody*
A	anti-A (α)
B	anti-B (β)
AB	none
O	anti-A and anti-B (α, β)

The selective advantages which encouraged either development of this antigenic system, or of immune responses to those antigens not present, are unknown. The system in its developed form is recent, and only found in the great apes. However, since substances with anti-A specificity may be found in beans and snails, and A-like antigens are found in pigs, the basic system is probably very old and may be part of a large complex.

The clinical importance is that fatal consequences may follow giving more than a few ounces of blood whose cells are agglutinated by the plasma of the donor. (Plasma is the natural fluid vehicle of the blood: serum is plasma from which various proteins are removed by allowing it to clot, after which the clot retracts leaving the serum). Solid organs grafted from persons who would be unsuitable as blood donors are also rejected rapidly. The giving of incompatible plasma is usually harmless due to dilution, and is sometimes necessary due to shortage of the rare AB type, so that AB individuals may have to be given A-type blood.

In emergencies group O negative blood is sometimes used without testing. Obstetric flying squads always take two pints for instant use in severe bleeding.

The ABO system rarely produces serious consequences for the fetus, partly because the main immunoglobulin class, IgM, is too large to traverse the placenta. Incompatible fetuses (that is, those which could not act as donors to their mothers), have a slight increase in the normal jaundice of the newborn, and have the considerable advantage that due to the rapid destruction of any fetal cells which enter the mother, they are unlikely to damage their sibs by the production of rhesus antibodies

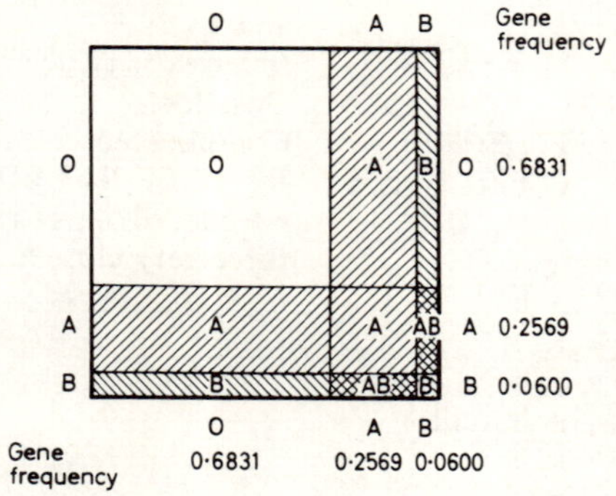

Fig. 4.3 The relationship of allele proportions (O, A and B) to phenotypes (O, A, B and AB). The areas are proportional to the numbers.

should they be Rh+ and their mother Rh–. (See below) In Europeans the A, B and O alleles have proportions of 0.25, 0.05 and 0.7 approximately and these lead to the following proportions of the phenotypes, A, B, AB and O (Fig. 4.3). In Asians the B allele is relatively commoner. Oceanic Europeans and Celts seem to have a higher proportion of group O. The American Indian is almost entirely group O, and may have been entirely group O before the colonization of the New World.

4.5 The Rhesus system

This was discovered by Landsteiner and Stetson in 1939, following their observation that a fetus stillborn with 'hydrops fetalis' had immunised its mother against her husband whose blood was being used for transfusion. At that time 'bottled blood' was rarely used, transfusion being direct, usually from relatives. Landsteiner and Weiner later identified the antibody as similar to one they had produced by giving blood from rhesus monkeys to rabbits, and the system and the disease it produced took this name. The antibody is now known not to be identical so the name is erroneous, but too well established to change. Thirty years after the recognition of this common genetic disease, which had been made considerably commoner by the use of transfusion, the spread of operative obstetrics and, in some countries, the injection of parental blood into sickly children, the disease was brought under control by immunological means and is now becoming rare.

The system behaves as though due to three closely linked pairs of alleles C,c; D,d; and E,e. Only D and d are sufficiently important to feature in routine transfusion typing. Antisera are available against all antigens except *d*, which behaves as though it were a null antigen, as the *O* in the ABO system. The proportion of the alleles D and d are 0.6 and 0.4, so that the phenotypes DD, Dd and dd have the proportions 0.6^2, $[(0.6 \times 0.4) + (0.4 \times 0.6)]$, and 0.4^2. That is, about 16% of people are

Table 4.3 Table of common blood group variants in man

System		*Chromosome*	*Alleles*	*Comments*
ABO		9	A, B, O	A consists of A_1 and A_2
MNSs		?	M, N; S, s	Behaves as two very close loci,
P		?	P_1, P_2	Complicated
Rhesus	Rh	1	C, c; D, d; E, e	R_1 = CDe, R_2 = cDE, r = cde. Behaves as three very close loci.
Lutheran	Lu	?	a, b	
Kell	K	?	K, k	
Lewis	Le	?	a, b	
Duffy	Fy	1	a, b, null	
Kidd	Jk	?	a, b	
Xg		X	a	

Rhesus negative. The classes DD and Dd are known as D+ or Rh+, dd as D– or Rh–.

If D+ cells are given to a D– person, either naturally in labour or following operative obstetric procedures, or by transfusion, antibodies may be formed which will damage any red cells infused later. A more serious consequence is that these antibodies are largely dimers (IgG) and small enough to get into any subseqent fetus and to destroy its cells. This can be completely prevented by giving anti-D to all rhesus negative women at the birth of their children. This causes any fetal cells which have strayed into the circulation to be destroyed before they have provoked much antibody formation. The amount of fetal bleeding can be estimated from a blood film in which adult haemoglobin has been leached by an acid buffer, so that fetal cells stand out; this is called the Kleihauer test. Antibodies can also be provoked by induced abortion and amniocentesis, and anti-D should then be given to all rhesus negative women.

Once antibodies are established any rhesus positive fetus will be at risk; the destruction of red cells leads to anaemia during pregnancy and severe jaundice immediately after birth, with a risk of brain damage from the high levels of bilirubin. If the baby survives birth, exchange transfusion with Rh. negative blood is usually curative. The baby may be transfused before birth, but this is difficult. This serious genetic disorder is now rare: indeed its rarity is imposing practical problems in the supply of the antibodies by which it may be prevented.

4.6 Other blood group systems

These are listed in Table 4.2. None present common clinical problems.

4.7 The human leucocyte antigen system (HLA system)

This will be discussed later under Disorders of Defence (Chapter 6). The system involves a segment of chromosome and includes several loci involved in recognition, and several others involved in the killing of invaders by the complement system. The A and B loci, now termed HLA-A and HLA-B, (formerly loci 1 and 2 or LA and FOUR, of the HL-A system) contain numerous alleles and, since they are close together, are usually undisturbed by meiosis. If one gamete conveys A1 and B5 alleles and the other A2 and B7, then only about once in a hundred times will there be a recombination, leading to gametes A1-B7 and A2-B5. This slow re-assortment, with a half life of about 2000 years, on a system undoubtedly exposed to strong and probably variable selection, leads to a persistent allelic association, some pairs of alleles, such as A1 and B8, being several fold commoner than would be expected if they were independently inherited. The excess proportion is termed Delta, written Δ or D, and is defined as the proportion found less than the proportion expected if unassociated.

Fig. 4.4 shows the distribution of A and B specifities from Birmingham. The term 'haplotype' is used for the set of alleles conveyed

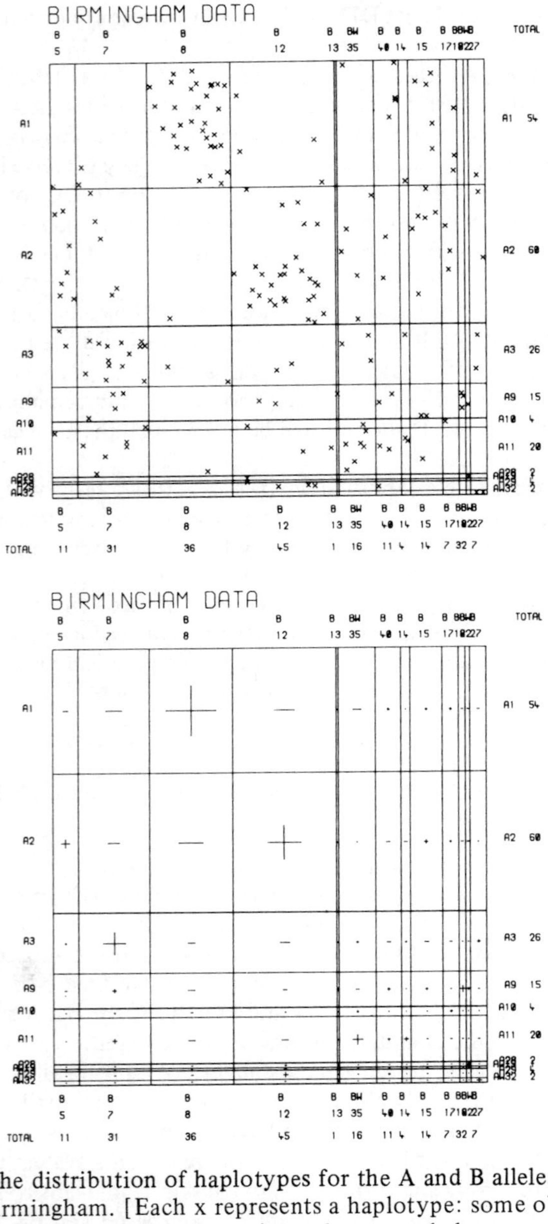

Fig. 4.4 The distribution of haplotypes for the A and B alleles, based on data for Birmingham. [Each x represents a haplotype: some obvious associations (A1, B8; A2, B12, etc.) are shown, and also some dissociations (A2, B8; A1, B12; A9, B18).] In the absence of associations these crosses, which are randomly drawn in the rectangle, would be evenly distributed. This is summarized in the lower rectangle, where the diameters of the + and the – are proportional to the excess or deficiency.

Table 4.4 The distribution of pairs of alleles at the A and B loci.

HLA-A		HLA-B	HLA-C	HLA-D	HLA-DR
A 1	B 5	BW42	CW 1	DW 1	DRW1
A 2	B 7	BW44	CW 2	DW 2	DRW2
A 3	B 8	BW45	CW 3	DW 3	DRW3
A 9	B 12	BW46	CW 4	DW 4	DRW4
A 10	B 13	BW47	CW 5	DW 5	DRW5
A 11	B 14	BW48	CW 6	DW 6	DRW6
AW19	B 15	BW49		DW 7	DRW7
AW23	BW16	BW50		DW 8	
AW24	B 17	BW51		DW 9	
A 25	B 18	BW52		DW10	
A 26	BW21	BW53		DW11	
A 28	BW22	BW54			
A 29	B 27				
AW30	BW35				
AW31	B 37				
AW32	BW38				
AW33	BW39				
AW34	B 40				
AW36	BW41				
AW43					

together in the gamete or haploid cell. The + and – refer to the excess or deficiency of any pairs, the size of the + or – being proportional to D. The new (September, 1977) nomenclature for the receptor loci and alleles is given in Table 4.4.

5 General inheritance

In the mendelian disorders there is usually no problem in defining the boundaries between normal and abnormal variation, since most of us would rather not be carriers for the offending alleles. Occasionally there is good reason to regard the carriers as having some advantage, known or unknown, but unless the malarious or other relevant hostile environments of the past return, such alleles can be regarded as abnormal for practical purposes.

In those disorders related to general inheritance, which is just another way of saying that all people differ, and some are more susceptible than others to disease, there are serious difficulties in terminology. The words multifactorial or polygenic are sometimes used, but these mean little and are best avoided. We start from the obvious fact that we are born of various sexes, sizes, and shapes and respond differently to exposure to the diverse environments imposed by food, bacteria, parents, teachers, housing and climate. Some of these differences, such as that of sex, are of a simple genetic nature and associated with large differences in disease experience. Some disorders, such as breast cancer or silicosis from coal mining, show a major difference in incidence. Most others, including disorders of childhood, or of such apparently sexless organs as the cerebellum or the colon, also show considerable differences. We could state that sex was a genetic factor influencing disease, but this would not be a very helpful statement unless related in a meaningful way to mechanism or opportunity.

More subtle genetic differences, such as blood group or HLA type, may also influence susceptibility to disease, but we can no more regard this predisposing factor as a disease gene, then we can regard the Y-chromosome as a disease chromosome due to its association with such disorders as some tumours, silicosis, or even criminality. Disease is a breakdown in health, which may be related to intrinsic wear-and-tear mechanisms, to invasion by parasites, to exposure to harmful substances, to a mutinous or anarchic response of the defense mechanisms, or to many other factors.

The response to the environment will necessarily vary with the inborn constitution, which will show a familial tendency, and this will be compounded inextricably by the familial tendencies in environment, including education and housing.

The intensity of familiarity may be measured in various ways, of which the simplest is the relative risks of brothers or sisters compared to the general population, a ratio termed k by Penrose. Far more complicated methods are also available, some of which prejudge the issue by using the word heritability when what is meant is familiarity. However, these methods involve serious theoretical problems and cannot yet be said to have led to any useful discoveries in man.

It is possible to state that, if a predisposition to a disorder is the result of a large number of genetic units, then, if the environment were uniform, Penrose's k would be approximately the square root of the reciprocal of the incidence in the population. That is, if the disorder were found in 1/900, k would be about 30. However, it is in no way possible to infer that, because k is of this value, therefore the determinants are numerous, or that the relevant environment is uniform, or that some environmental factor is not essential. Rickets, silicosis, haemophilia and road accidents to children all show large familial tendencies. If, for example, we were to study pigmentation and sunburn, where we have a basically simple and variable inborn protection against sunlight, we can immediately see the problems which would beset an army of blind dermatologists, well armed with statistical advice, who set about studying the aetiology of this strange familial and seasonal disorder.

We could define among the victims some rare cases of congenital lack of pigment, the albinos, who were also handicapped by poor eyesight, and whose parents could be advised that their further children would be highly at risk. The same could be said of the more numerous groups of redheaded children. Most victims however would merely be fair, and many would be victims of foolish exposure, of occupational necessity, or of an inability, from poverty or vanity, to clothe themselves adequately.

Most poorly understood disorders in man are at least as complicated as this disorder would be to the blind, and diseases are becoming more complicated, since those whose mechanisms are simple are the first to be understood and prevented.

5.1 The familial pattern of general inheritance

Mendelian disorders are characterized by a high incidence in appropriate relatives, this incidence being independent of the severity of the disease, or the previous incidence in relatives. If, for example, the legitimate child of a couple has albinism, or phenylketonuria, the recurrence risk to further children is a quarter. If they then have another affected child, the risk is still a quarter, and this risk is unrelated to the severity of the disorder. If, on the other hand, a child of an unknown couple is very fair, dark, tall, intelligent, or stupid then we may infer that further children are likely to be similar, and if two children are distinguished in the same way, the third is even more likely to be similar. Where the variable considered is developmental punctuality, and the penalty for unpunctuality is a malformation, the same general tendencies persist.

In practice the recurrence risks are similar to those we would expect from a *variate* determined in such a way that the correlations were about ½ in the first degree relatives, such as brothers, sisters, parents or children, and ¼ in second degree relatives (who are the more distant first degree relatives of first degree relatives). However, while providing a rough guideline for advising on the recurrence risks of malformations this gives no guidance on whether this familiarity is increased by a familial environment, or whether the disorder is dependent on some extrinsic factor such as a teratogen or an infection. In general the recurrence risk drops off sharply with increasingly distant relationships, the logarithm of the incidence being approximately proportional to the degree of relationship. For practical purposes the empirical risks of most focal malformations vary between 1/50 and 1/20, and usually the only increased risk is of the same malformation. By focal malformation is meant one that could plausibly be related to a single developmental error. In generalized malformations, sometimes called multiple malformations, the recurrence risk is dependant on the cause; it is usually lower than in the focal malformations.

This is a field in which elaborate numerical or verbal formalizations seem to have little to add to the classical methods of natural history adequately supported by an understanding of the general nature of inborn variability.

The problems generated are not of a purely medical nature. Profound decisions relating to education are being made as a result of the poorly understood implications of studies of doubtful relevance to the matters at issue. The subject was well summarized in a recent leader in the Times which followed a remarkable series of letters, many of which implied that the brain differed from the body in having no inborn tendency to parental resemblance; this is reprinted below with permission.

The origins of intelligence

'The controversy over the inheritability of intelligence has made extremely interesting reading, though it has become more and more doubtful, as the controversy has gone on, whether it is the evidence, the definition of intelligence or the question itself which is the most unsatisfactory.

Without denying the possibility of the human mind having a spiritual dimension, it is clear that the human brain is a physical object which operates by physical processes. Its function is that of communication and control; in order to operate effectively it has to be able to perform a number of different tasks. As with a computer, those tasks can be simplified into input, of the acceptance of incoming signals, data processing, the sorting and retrieval of memory, and output. A very simple example of this scheme of the human brain might be the recognition and greeting of a friend in the street. There is a visual input as the friend is seen; the visual data is stored and compared with memory so that the friend is recognized; the response is decided; a greeting is spoken and a friendly visual expression is adopted – that is the output.

Even with that simple example it is certain that both inherited and environmentally determined mental activity is involved. We inherit physical equipment which enables us to see, to interpret sight, to remember and to learn to talk. We acquire the information which enables us to recognize, to remember and to speak a language. We develop by practice the brain tracks which facilitate recognition and speech. We are born with the capacity, we acquire the information, we learn to use brain functions.

It is clear that all brain activity, other than reflex or instinctive activity, requires the existence of two elements. The first is that the brain itself must be capable of the activity. No two brains are physically identical and it is safe to assume that no two brains have precisely the same capacity. The second requirement is that the brain should be programmed and have the data to perform the activity; it should have both the training and the information.

The physical brain, like the physical nature of any part of the body, must have its maximum potential entirely determined by its genetic inheritance. Of course maximum performance may never be reached for environmental reasons, just as a gazelle in a zoo will never run as fast as a gazelle in the wild. But all physical potential is a matter of inheritance, and brain potential is in this respect no different from muscle potential. The development of that potential, and therefore the development of anything that can usefully be termed intelligence, is almost entirely a matter of environment, of external influence. A sufficiently deprived environment will destroy any degree of potential; sufficiently low potential cannot be developed by however favourable an environment. The two processes are totally interdependent and intelligence depends not on a mixture of the two but on environment acting upon inherited potential.

The brain is much more complicated than any computer as yet designed and much less perfectly understood. As intelligent thought is inevitably dependent both on the physical inheritance of the brain and on the multiple inputs from the environment over the whole lifetime of the individual, the attempt to determine the relative responsibility of inheritance or environment must fail.

The only statement that can safely be made is that high intelligence will depend upon a conjunction of a favourable genetic inheritance and a favourable environment, though we do not know enough to be able to predict what will be a favourable genetic inheritance nor do we know enough to be able to say what will constitute a favourable environment, except in very general terms. To ask whether intelligence is the product of inheritance or environment is not like asking whether a whisky and soda contains more whisky or more soda, it is like asking in what proportion steam is produced by water or by heat.'

The simplest model of general inheritance is the threshold model of Pearson (Fig. 5.1). If, for example, we regard some distribution such as that of numbers of people with different blood pressure, then some threshold would define 'hypertension'; if the underlying variation were familial, then brothers and sisters of persons who were 'hypertensive' would be expected to have a distribution whose mean was between that of the normal population and the hypertensive population. Similarly, if we define tallness as being three inches above the mean for persons of the same sex, we could define pairs of fathers and sons by a point, and enumerate the various pairs tall:tall, not-tall:tall, tall:not-tall and not-tall:not-tall (Fig. 5.2).

Provided that these measurements fall into a fairly regular distribution, with one central 'hump', where most pairs lie, and that this is symmetrical about the main diagonal, then the distribution of these points can be summarized by the correlation coefficient of Karl Pearson, usually known as r. If we do not have the measurements, but only the categories defined by tallness, we may also estimate r from the numbers

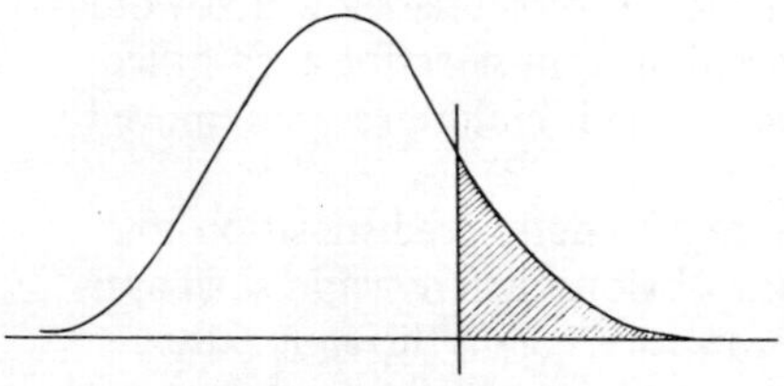

Fig. 5.1 Pearson's univariate model.

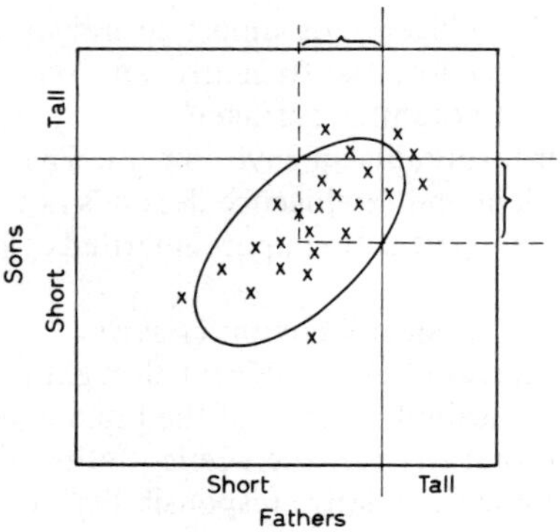

Fig. 5.2 Pearson's bivariate model.

by methods which are exact and complicated, or which are adequate and simpler.

Furthermore, if we postulate that a measurement is the result of a large number of genetic units of small and independent effect, and that the environment has either no influence, or only a consistent influence, we may compute the expected value of *r*. This will be a half for first degree relatives (parents and children, or brothers and sisters) and $(\frac{1}{2})^n$ for *n*th degree relatives. Uncles and nieces, and grandchildren and grandfathers, for example are second degree, first cousins third degree.

Obviously if the environment is important and erratic it will have the same effect as measuring with an erratically inaccurate machine, and weaken the value of *r*. If the environments vary between families *r* will confound environmental and inborn determinants.

In practice *r* measures familiarity, and familiarity cannot be broken into its genetical and environmental component without making assumptions of both independence of mechanism and independence of effect. That is, familiarity may be estimated in man, but its genetical component, known as heritability, cannot usually be defined on the type of data which can be acquired in a species intolerant of planned and randomized breeding and adoption programs.

We must assume, *a priori*, that since the body and the brain are made of cells, and substantially formed by birth, the correlations of height, weight, muscular strength, or mental ability are likely to be about 0.5 between sibs, or between parent and child, in a uniform environment.

All the many measurements made on body and mind in relatives are consistent with this. However, due to the familial correlation of both good and bad nutrition, or facilities for exercising the body and the mind, it is not possible to estimate the degree of similarity which would persist in relatives if parents were prevented from selecting each other for physical and mental similarities, and if their children were removed and randomly exchanged at birth.

Perhaps the most important evidence of genetic predisposition to physical and intellectual aptitude is the obvious difference in such aptitudes and achievements in brothers or in sisters. This difference is so marked that it is obvious to anyone on casual observation. Those who wish for documentation may note the rarity of sibs who reach similar distinction in athletics, music, chess or mathematics, or even such complex behavioural aptitudes as those involved in politics, the theatre, or even sophisticated crime.

6 Disorders of defence

The defensive system of the body takes up a volume, and has an energy requirement, comparable with that of the nervous system; most of its activities involve transfer and storage of information. Although the mechanism of a system almost devoid of rigidly connected and non-reproducing cells is very different from that of the brain, the amount of information transferred and retained may be comparable.

Such a system has evolved in response to the competitive interests of protozoa, bacteria and viruses, and these organisms can both out-number man and individually outbreed him with a ratio of generation times of minutes to years. The gut of one man has a bacterial population greatly exceeding the human population of the world, and many bacteria can double their numbers hourly or faster. Against such massed numbers, with new mutants constantly arising to exploit any weakness, mammals have had to deploy a system which maintains the ability both to respond in days, and to remember such responses against further attack. Furthermore, the species will obviously benefit from any mechanism which will allow survivors from one epidemic to have the genetic background which assisted their survival retained for possible future use.

The two major systems which have evolved are that related to the censorship or immigration of molecules and organisms, and that related to the recognition, inactivation and destruction of organisms which have gained entry. The former system, which is presumably the more primitive, may be termed the cellular system and the latter, which is better known since it forms the basis of most artificial methods of improving immunity, as the humoral system. The humoral system is, of course, manifested by cellular activities and secretions. Nevertheless, the distinction is useful, and the more exact 'cell membrane defence system' a little cumbrous.

The cellular system seems largely related to a wealth of specific receptors which are so complex and so numerous that, where some receptors have been deceived into allowing the entry or adhesion of invading organisms, there will almost always be some members of any population surviving. So far as is known these receptors are inborn and not 'educable' by previous experience. The most complicated receptor system known is that related to the HLA system (Human Leucocyte Antigen) (see p. 00), but many blood group systems seem to be receptor systems whose primary aim is the deployment of diversity in the membrane of some cells.

The humoral system is largely manifested through the synthesis of molecules, the light and heavy immunoglobulin chains, which exhibit variability both from true genetic variation and from somatic variation which is acquired in response to the differential survival of responding cell lines after conception and throughout life.

6.1 The humoral system

Although presumably not the most primitive system, since it is largely related to the synthesis of antibacterial agents in the blood stream, and defence mechanisms certainly preceded the development of extra-cellular fluids, it is the best understood system, largely because blood is easier to study than the cell membrane. The effective molecule is one of several varieties of immunoglobulin, which are compounded into polymers, and whose units consist of heavy and light chains in the general form:–

```
l  –  L
      |
h  –  HHH
        |
h  –  HHH
      |
l  –  L
```

when the small letters refer to the variable regions and the capital letters the constant regions. Some heavy chains, including those in immunoglobulin E, have a fifth unit. The constant regions may share variation between individuals, and these variants form the Gm, Am and Km (formerly Inv) allelic types or allelotypes and are distinguished by single amino-acid differences. The variable region is consequent upon some poorly understood event which leads to an effective high mutation rate in the somatic tissues. The 'teeth' of the molecule are between the *l* and the *h*. On exposure to any antigen those cells which are producing molecules whose 'teeth' will grip exactly the antigen are in some way encouraged to reproduce, so that on a second exposure the antigen will receive a vigorous reception. This is the basis of the regular immunization procedures used to protect against typhoid and tetanus. Recognition is not a fatal event for the invader; once recognized, various killer systems, of which the complement system is the best understood, must be mobilized to destroy any cell recognized, rightly or wrongly, by the recognition system.

Such a defence system may be a cause of disease if it fails to defend, or if it attacks the host rather than the invader, or if it discriminates against something essentially beneficial, such as a blood transfusion, or essentially harmless, such as a few drops of the baby's blood at child-birth. Furthermore, some organisms, such as edible plants, may protect themselves by disrupting or diverting the defence mechanisms so that they lead to a response which is itself disabling, as in hay fever and asthma. As in other diseases we may break down the classification into simple mendelian disorders and into disorders due to general effects or predispositions.

A failure to synthesize some component in a recognition system, as in X-linked agammaglobulinaemia, in which there is gross deficiency of gammaglobulin, is a simple example. A failure to follow up recognition by execution may also arise, as in some defects in the complement system. Finally, the sheer complexity of the system exposes it to spontaneous mutinous or neoplastic activity, as in the autoimmune diseases and some lymphomas and myelomas. The latter are of peculiar interest, since they are usually clonal tumors derived from cells so specialised that they produce only one variety of both light and heavy chains, and any natural or genetic variability has been suppressed due to only one of each pair of alleles being active.

All the known units in this complicated process are very variable and differing units doubtless lead to differing responses for good or ill. As with a military defence system, the optimal proportion of the budget alocated, the tolerance of casualties during training and practice, the advantages and disadvantages of diversity and standardisation, or of speed and sureness, are difficult but necessary decisions. A completely reliable system would be too slow, and a completely standardized system would be unable to respond to the unexpected. An exhaustively tested system would be too late. In man the defence system has evolved by survival and not by planning and it has been tested against the enemies of the past, mainly the remote past, and has no necessary defences against completely novel environments, In some situations, as in blood transfusion, transplantation, or the replacement of insulin or clotting factors by injection, the defence system may itself be a major cause of difficulty. Hayfever, asthma and eczema are common and irritating consequences of excessive intolerance of otherwise harmless proteins.

6.2 The cellular defence system

This is too poorly understood to be discussed in detail in an elementary book. In the HLA system (the human leucocyte antigen system), determinants of a large number of receptors are recognized at several loci (HLA-A, -B, -C. -D and DR), and these are determined by closely linked loci, between which are at least three elements of the complement killer system, including both the classical pathway components C2 and C4 and the alternative pathway component Bf (B factor). Several disorders, including a pathological response to gluten (coeliac disease), an extremely unpleasant response to some poorly defined infections (ankylosing spondylitis and Reiter's disease: the latter was previously

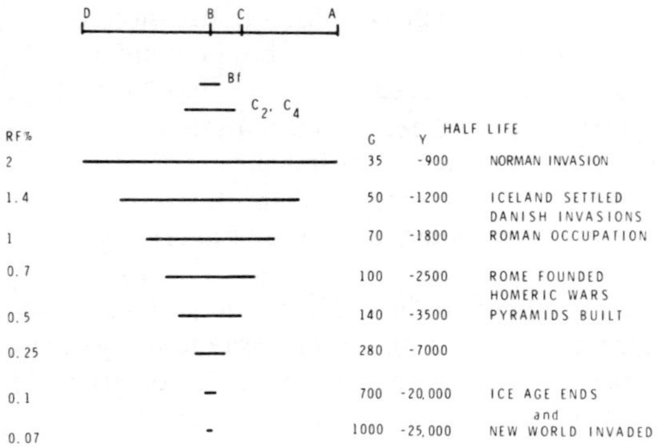

Fig. 6.1 The approximate positions of the A, B, C and D loci, and their distances in terms of recombination fractions (RF) and half life in generations (G) and years (Y).

thought to be venereal in origin), myasthenia gravis and diabetes, are known to be strongly influenced by the inborn HLA phenotype.

In view of these loci being so close the alleles at one locus will tend to be associated with the alleles at another, since rearrangement by crossing over at reproduction will be rare. (The A and B loci are rearranged about once in a hundred births, so that about half the segments of the chromosome involved will go back unbroken to the Roman occupation of Britain: Fig. 6.1.) Consequently apparent associations at one locus may reflect those at a neighbouring locus.

The only interpretation which can be placed on these disease associations is that they suggest the importance of some agent whose entry, or the response to whose entry, is influenced by receptors. The most obvious agents are plants trying to protect themselves from being eaten by the use of toxins, and viruses trying to enter or leave cells. Most plants and all viruses are more numerous than man, and breed faster, so that their systems of defence or subterfuge are likely to have changed in the centuries or millenia since they encouraged the selection of the commoner determinants of these receptors.

Such associations cannot meaningfully be interpreted in terms of 'disease genes'. Our ancestors have run the gauntlet of the past and their survival depended upon the variability of individual responses to parasites and to foodstuffs. If some alleles, such as those predisposing to coeliac disease or diabetes, do not seem to have any mitigating advantages now, at least they must have had in the past. In view of the tight linkage the alleles are associated with one another, and the association of one locus with disease may be secondary to a neighbouring locus which may be functionally unrelated.

6.3 The malarial defence system

Malaria has been a major limitation on human activity in some of the most fertile regions, and, since the parasite spends part of its life cycle living in the red cell, and eating its contents, the defence mechanisms known include those related to entry to the red cell, and to survival after

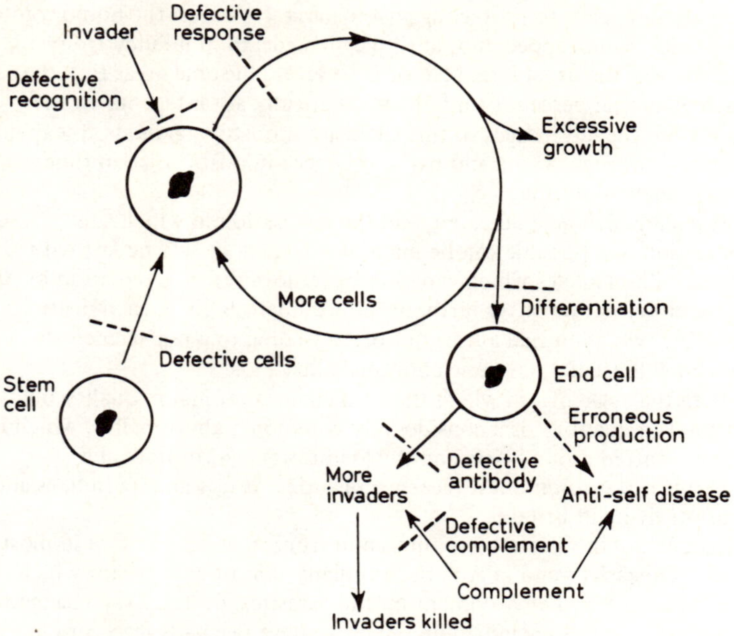

Fig. 6.2 A simplified version of one type of defence and of the various disorders which can disrupt it.

entry. These defence systems are responsible for the commonest lethal variants in man, and are maintained at a cost of many millions of variant alleles, and of tens of thousands of deaths a year, from thalassaemia and sickle cell disease.

The cell entry defence system has only recently been described. The parasite appears to need the receptors of the Duffy blood group system to gain entry to the red cell. A few people lack the commonest receptors (Fy^a, Fy^b), and may lack any receptor, and the parasite is unable to gain entry. The other disadvantages, if any, of lacking these receptors are unknown.

The internal cellular defence system is more subtle, and works largely through the preferential destruction of those cells which are parasited.

Normal adult haemoglobin is a tetramer composed of four units, two haemoglobin alpha chains and two haemoglobin beta chains. In sickling a single amino-acid change (glutamic to valine), can be fatal, and is caused by a single change in one of the 3 000 000 000 nucleotides in the haploid set. As a result when cells containing this haemoglobin are exposed to reduced oxygen tension the molecules tend to adhere and distort the cells into odd shapes, including crescents or sickle shaped cells. These cells are preferentially destroyed in the spleen, and also tend to jam when compacted in the narrower blood vessels.

In the heterozygote the presence of this abnormal beta chain has little effect except to protect against malaria. However, the homozygote is seriously handicapped and, until recently, death in infancy from anaemia was the usual fate. Malaria is so severe in some areas that the death of several percent of infants is genetically advantageous to the group, and the advantages of this allele are probably adequate to explain its spread through Africa and parts of Greece in post-Christian times from a single mutation.

If malaria can be eradicated, and the civilisation in which this eradication was possible can be maintained, then we will be left with several million alleles which can only be removed from circulation by death, either natural after birth, or induced before birth, each death removing two abnormal alleles but often leading to a replacement by a sib who will usually have one abnormal allele.

Beta thalassaemia, in which the beta chain is normal in quality but deficient in quantity, is a considerably commoner abnormality, which is mainly centred around Asia and the Mediterranean. In Italy and Greece it is the commonest recessive disorder, as it is among Indians and Cypriots living in Britain.

Haemoglobin variation is common in tropical areas, but rare in most Nordic Caucasians and in American Indians; almost any variant which will impose a novel environment on the parasites, or delay the changeover from fetal to adult haemoglobin, seems to have been advantageous in malarial areas in the past.

The red cell is very prone to parasitism, especially as it provides the normal food of blood-sucking insects which can act as vectors. It is also fairly easy to study, and, as cells go, very simple. It seems likely that malaria and red-cell variability is typical of the relationship of genetic variation and many diseases.

Although the basic mechanism seems simple, such questions as 'is the sickle cell allele a normal or pathological variant?' cannot be answered without reference to the environment. In the presence of malaria it is a normal variant, imposing health on some and death on others. In the absence of malaria it seems unconditionally harmful for as long as malaria remains absent.

7 Other disorders

The neoplastic disorders, like the infective disorders, are largely the result of a failure to tolerate various environmental agents, in this case chemicals which can impose a spontaneous 'infective' behaviour on the cells on our own bodies. No doubt selection is slowly removing from potential parentage these few victims afflicted while young, but such selection is likely to be very slow, and many millenia would be needed to halve the risks to the smoker.

Generalized disorders, that is, disorders not known to be related in any simple way to any single genetic or environmental factor, make up the bulk of disease, and will provide the main exit from life for most of us. Like all disorders they are familial, usually fairly weakly so, and most of us will have about double the risk of suffering from the disorders of our parents, and of following the same exits from life. Most of these disorders, which include coronary artery disease, strokes, high blood pressure, and diabetes have a steady, and usually exponential, increase with age and are strongly influenced by such private environments as those related to food, exercise, smoke and excessive drink; many can be greatly mitigated by therapy. They are sometimes termed multifactorial diseases.

In the congenital malformations, which make a more tragic impact than even the early deaths of those who were once fully formed, there are likely to be familial differences in the tolerance of teratogens, and, even in the most perfect environment, anything so complicated as an embryo is likely to go wrong in many ways; these ways are likely to show familial tendencies. However, those predisposed can usually only have this predisposition made made manifest by the event, and, since the tendency to recur is usually fairly small, neither natural selection, nor the artificial selection following a decision not to have further children, is likely to have much influence on the predispositions of the succeeding generation.

So, while the effect of selection in a constant environment is to discourage the perpetuation of any predisposition to death in children and young adults, the fact that death is inevitable, and, indeed, if timely, has a selective advantage in most societies, and that most mechanisms of death are intrinsically liable to unpunctuality, means that after many millenia of adaptation, we, as a species, have a genetic variability which is reflected in both health and disease. Susceptibility to most disease is not readily amenable to artificial selection, even if it were expedient to attempt to impose such selection. A small minority of disorders are necessarily related to specific genetic units, and, in these, artificial selection has some prospect of reducing suffering, as has the preemption of birth by death by fetal diagnosis and selective abortion.

Conditions not due to one genetic unit must necessarily be due to more than one, and disorders related to several or many are usually also greatly influenced by the environment. These are sometimes termed multifactorial or polygenic; however, such words give a misleading impression of knowledge, and are conveniently avoided, since they cover all disorders other than that minority due to single factors.

It is, however, useful to distinguish disorders related to a necessary and known extrinsic factor, as, for example, infections, lung cancer and sunburn, and those in which no necessary factor is known. This latter group, of course, includes those in which there is an unknown necessary factor, and those in which there is no necessary factor.

All disorders could be made rarer by selective breeding, that is, by artificial selection, but the effects will be slight. In diabetes, for example, 5% of children with diabetes have a diabetic parent, and even if these parents had not had children, then over 95% of diabetic children would still be with us.

Recurrence risks for fetal disorders have been tabulated accurately, and consistent data are available from many cities, especially London, Montreal and Birmingham, and the parental problem 'will it happen again', can usually be answered with precision. In practice, most malformations seem to recur at a rate of about 3% or less, and they 'breed true'. Disorders of later life have a similar familiarity.

Information on such risks may influence subsequent conception rates, but the effects will necessarily be slight, since most cases are first cases so far as the family is concerned.

7.1 Mental disease

The major categories of mental disorder are those of inadequacy of response, which may be congenital (amentia) or aquired (dementia) and those of inappropriateness of response (the psychoses). Psychoses are divided into the major catergories of disturbances involving excesses of depression or elation (mania), the manic-depressive group, (and many victims oscillate wildly from one to the other), and disturbances of concept formation, the schizophrenias, which include both organized disturbances, as in coordinated delusions of persecution, and chaotically disordered thought.

These various disorders, all of which are merely extreme forms of behaviour which are compatable with normal life in mild forms, tend to run in families, and tend to 'breed true'.

Social isolation, or physical segregation, will reduce the potential to breed, and, as such, reduce the incidence in the next generation. Unfortunately, the benefits of this are far from obvious, due to the great contributions which psychotics often make before their madness. The drive of the slightly manic, the deep and prolonged cerebrations of the depressed, and the imaginative capacity of the dreamer, could not readily be distributed evenly, and, it may be that the benefits such individuals confer on society can only be harvested if society supports those seriously disabled by peculiarities which are beyond the tolerance of employers or relatives.

Those whose primary problem is too little mental activity, rather than inappropriate activity, are a far graver burden since children necessarily tend to resemble their parents, and social planning which aims to 'rehabilitate the mentally handicapped in society' cannot easily be achieved without rehabilitation including reproduction. Little consideration has been given to the genetic consequences of such policies, partly due to a belief that disorders are genetic or non-genetic, and partly due to systematic misunderstandings on the nature of regression, a rather complicated statistical concept. (See Further Topics, Chapter 10.)

In practice 'genetic' mental deficiency is usually recessive and any children are likely to be genetically normal, while 'regular' or 'unclassified' mental deficiency is likely to be mere dimness, and the children are likely to have an intellectual capacity intermediate between that of the parents. Those with unusual mental aptitudes, either high or low, tend to prefer the company of their peers.

There are no practical policies for the discouraging of the reproduction of those least able to either beget children with average potential, or to rear them in an adequate environment, beyond the provision of good institutional facilities within which adequate happiness can be achieved in a disciplined and segregated environment. Such private measures as genetic advice to those requesting it can have no impact on mental deficiency, and such public measures as sterilization cannot be carried out in a society worth living in on those who do not understand what is involved.

7.2 Neoplastic disorders

By neoplasia is meant 'new growth' or 'cancer'. Most cancers are of the epithelial cells, mainly those lining the gut, lung, and the urinary and reproductive systems, and are known as 'carcinomata'. The rarer mesodermal tumours are known as sarcomata. Leukaemias, Hodgkin's disease, and lymphomas are difficult to classify, and not covered by either carcinoma or sarcoma.

Various rare recessive disorders involving a failure of DNA repair are now known, and these usually lead to tumours of the 'protective' tissues, the lymph nodes and the parts of the bone marrow producing white cells. There is also some slight increase in predisposition in carriers. However, these rare conditions are responsible for less than 5% of such tumours.

Rare dominant disorders which lead to tumours also occur, the commonest of them being retinoblastoma, a tumour which is effectively treated by removing the eyes, The inherited forms are, as is to be expected, usually bilateral, and necessarily becoming commoner due to treatment.

7.3 Neoplasia

The various mechanisms by which tumour predisposition may be manifest are shown in Table 7.1.

Neoplasia is the failure of a cell to limit its reproductive powers to the demands of the organism as a whole. All cells must be assumed to have a very small risk of going berserk in this way, and there are presumably mechanisms for removing from the body such dangerous cells. Some are remarkably safe: the red cells precursors are reproducing at about 10^6 a second, and yet related tumours are rare. The small gut, which

Table 7.1 Examples of tumours and neoplastic change related to the likely cause in various disorders, all of which are rare.

Number of mitotic errors =				
No of cells	x	*Mitotic rate*	x	*Error rate*
Polyposis coli Melanoma in gray horses		Traumatic tumours		Fanconi's syndrome Bloom's syndrome Some allied syndromes X-rays Sunlight Tobacco Melanoma in albinos

protects itself from digestion by a steady production of suicidal cells, is also rarely afflicted.

An increase in incidence occurs when there are more cells at risk, when the accuracy of their reproduction is impaired, when they are more exposed to disturbance, or when the systems of recognition and destruction are impaired by disease, or more usually, by its treatment.

Tumours seem widespread in most vertebrates, and take various forms. The wide variation in incidence in man by place, a variation quickly adopted by the migrant, suggests that most forms are related to various environmental agents, most of which are also mutagenic.

8 Artificial selection

Our present disease experience reflects the interaction of the genetic contribution, which is uniquely assembled from components which have survived the past, and of the environment within which we live. Different ancestral backgrounds have had different problems of survival, and their descendants will continue to have different problems in adaptation. In the majority of disorders there are considerable differences in susceptibility which are determined both by remote ancestry and by the exploitation of chance at meiosis, but these differences are rarely sufficient to burden those at risk with the added problems of adjusting their reproductive activities for the convenience of other people's descendants.

In those disorders which are simply determined by genetic mechanisms the risks of the hazards may be so high, and the consequences so unpleasant, that those at risk may decline to expose their descendants to such risk, or themselves to such distress.

Genetic disorders may, in principle, have the incidence reduced by

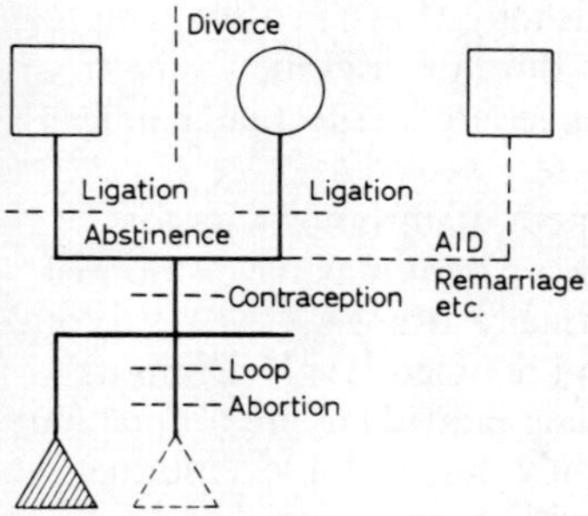

Fig. 8.1 Options available in genetic counselling.

influencing marriage, by obstructing conception, and by abortion, in some cases after diagnostic examination of the fetus, or of fetal cells or fluids.

These various approaches comprise artificial selection. The commonest situation follows the request of a parent who has an affected child for advice as to what is the chance of this happening again, and whether anything could have been done, or could be done, to mitigate or prevent such an occurrence. Occasionally advice may be asked before conception, or even before marriage, but such advice is relatively uncommon, and, in view of the limited circumstances in which such advice is indicated, it is likely to remain relatively uncommon in adequately informed communities.

Such situations, in which the advice is requested, we will term private. There is some confusion as to the demand for such advice, and also as to its efficacy. Many couples have a sufficiently strong desire for normal children to continue to wish to conceive after being advised of a high risk of recurrence, and there is no justification for tabulating such responses as failures, or for assuming that the births of second affected children reflect on the standards of either medical care or personal responsibility.

Most clinicians consider gratuitous advice on recurrence is indicated when the risks are high, say over 10%, and that gratuitous reassurance is indicated when they are low, say under 1%, and leave the patient, or one of their parents, to enquire about the prognosis in the intervening grey area. These matters are difficult, and there is no reason to regard uniformity as preferable to variety. We may summarise the rather obvious options in Fig. 8.1 and consider some specific problems which commonly arise in matters of a private nature, that is, when the referral relates to a specific problem within a family.

We will consider specific examples of diseases due to chromosomal imbalance and to dominant, recessive and X-linked disease.

8.1 Chromosomal disorders

The consequences of an additional chromosome vary from rather minor disturbances for an extra X in a female or an extra Y in a male, to very severe consequences from an additional autosome; the most serious in practice is trisomy 21, described as Mongolism by Down in 1866, and often known as Down's syndrome. About a third of cases die young from cardiac defects, the remainder surviving with severe intellectual handicap. The other trisomies rarely survive beyond a few weeks.

Imbalance arising as a result of a rearrangement of chromosomes, or translocation, are of particular interest since a parent may have a normal set of chromosomal material which is abnormally arranged, as a result of two normal chromosomes having broken and rejoined (see McDermott's *Cytogenetics of Man and other Animals*). Such individuals are normal, but any of their children who inherit one, and only one, of these abnormal chromosomes will have an unbalanced set, with deficiencies of some genetic material and excess or duplication of other, and severe development disturbance with generalized distortion in shape and severe intellectual handicap is usual. It is important to appreciate that the chromosomal defect in the parent will always be more obvious than in the affected child. In such cases diagnosis is in priciple possible *in utero* and recurrent cases are amenable to abortion. There is no indication for aborting the balanced form which, in practice, seems particularly common due to disturbances in segregation. The unbalanced forms seem to have an expectation of about 10% rather than the expected 25%. Further details may be found in Hamerton (1971; see Bibliography).

Although important, such conditions are rare, being exceeded in frequency by simple trisomy, mainly trisomy 21, by a factor of about ten. Translocations involving chromosome 21 and another acrocentric are the commonest variety, and usually arise *de novo* from normal parents.

In general the risk of recurrence when the parents have normal chromosomes is usually low, being about 1% in trisomy 21 and less in all other numerical disorders.

From the practical point of view all parental translocations which are known to have been associated with an affected child are an indication for fetal diagnosis in those wishing it; trisomy 21 has a recurrence risk about equal to the spontaneous risk at forty but, since such parents have personal experience of the condition, the indications for fetal diagnosis are usually stronger in view of the more pressing anxieties. In other trisomies, or in translocations with normal parents, or in mosaicism, the risks of recurrence are far lower and hardly justify fetal diagnosis outside highly specialized centres where the procedure itself has a lower risk than the condition.

Chromosomal disorders are obviously capable of diagnosis *in utero* although the prognosis from some of the milder forms, such as the XXX female and XYY male, is poorly known, and must await substantial surveys of adults at autopsy. The efficiency and effectiveness is purely a technical problem of the safe collection and reliable analysis of amniotic fluid on those who wish it done. At present, in most centres, the hazards of the procedure are probably between ½% and 1%, or equivalent to the incidence of serious chromosomal inbalance between thirty-five and forty years of age. There is evidence that paternal age also contributes, about one third of the additional chromosomes being paternal.

The efficiency is limited by the hazards of the procedure itself, which is likely to decline, and is probably as low as ¼% in some centres. The procedure is, in principle, completely effective.

8.2 Dominant disorders

These may either be inherited, or may be the result of new mutations. The latter cannot be predicted, although some prevention may be possible by reducing the use of mutagenic substances and procedures and by protecting the gonads on X-ray;

Inherited forms may, in principle, be prevented by those afflicted not reproducing: however, in practice, this is not usually acceptable. Furthermore, those afflicted may not be clear of their affliction until after reproducing. A classic and distressing example is Huntington's chorea, which has an incidence of potential affliction of about one in ten thousand at birth in Anglo-Saxons, and is higher in the coastal areas around the North Sea. It seems to be maintained almost entirely by transmission, mutation being very rare and, as yet, not clearly documented in any individual case. A proportion of those at risk would be sterilized if there were any reliable test for the disease. The efficiency of this is 50%, in that every two non-births will lead to one normal non-birth for each affected non-birth, but it is completely effective.

8.3 Recessive disorders

The incidence of recessive disorders is known to be about ¼% at birth, a few areas having an additional incidence of sickle cell disease or thalassaemia of up to a few percent.

If all the various disorders are considered, and the proportion of alleles which carry the defect is estimated, then it is clear that, on average, each gamete must convey between a half and one defective allele. Far higher figures abound in the secondary literature, but these are mainly based either on very indirect inferences or on limited or questionable data.

As an example fibrocystic disease, the commonest Angle-Saxon recessive, affects almost one child in 1600, who is homozygous. In the absence of inbreeding this means that, if we merely consider one-child families, since in a marriage between carriers only one child in four will have the disease, one marriage in 400 must involve carriers, and therefore one person in 20 must be a carrier.

That is, if the frequency is one in 1600 the incidence of carriers is one in 20, since a carrier could carry it at either his paternal or maternal loci. (The argument based on only children is simple and not biassed.) Alternatively, we may note the incidence is the square of the gene frequency ($1/1600$, is $(1/40)^2$), and the carrier frequency for rare conditions is twice the gene frequency.

A model of all possible two-child families (Table 8.1), shows that

Table 8.1

0 0	0 0	0 0	*	0
0 0	0 0	0 *	*	0
0 0	0 0	0 *	*	0
0 0	0 0	0 *	*	*

most cases will be the first in the family, and that the stopping of second affected births would reduce the number affected in the population but slightly. In Britain the family size structure is such that while the efficiency will be only 25%, the effectiveness, which is influenced by family size, will be about 15%.

A proportion of recessive disorders are manifest *in vitro* in cells cultured from the amniotic fluid, and in these cases fetal diagnosis is possible. In Britain the commonest conditions amenable to such diagnoses are the polysaccharidoses in which normal substances accumulated in abnormal amounts. Hurler's disease, or gargoylism, is the commonest in Britain, and Tay-Sachs' disease in Jewish populations. Fibrocystic disease cannot yet be diagnosed *in utero*, and it seems unlikely that many of the numerous disorders leading to deafness and blindness will be manifest in undifferentiated cells. Many disorders can now be diagnosed but most are of extreme rarity with an incidence of less than one per paediatrician lifetime (50 000–100 000 births).

Disorders of haemoglobin production (the thalassaemias and sickle cell cell disease) are the commonest disorders on a global scale, and their recognition *in utero* is limited by the dangers of sampling fetal blood and the technical problems of examining mixed bloods. Rapid progress is being made in both fields.

8.4 X-linked disorders

Duchenne's muscular dystrophy is the commonest X-linked lethal, and is maintained by mutation to such an extent that only a minority could be prevented from development by prevention or elimination. An unreliable test of carrier status is available by estimating the level in the blood of creatinephosphokinase, an enzyme which leaks from damaged muscle. Unfortunately, this is the best available test, and if estimated reliably, and evaluated against the pedigree, a proportion of potential carriers may be given fairly reliable advice. However, many will have intermediate values. The test is completely unreliable in pregnancy which lowers the value to an unknown extent. Fetal diagnosis is impossible, except that maleness can be determined by nuclear sexing and males aborted: half would have been affected and half of the surviving females will be carriers, almost all of whom will be completely unaffected. A few carrier females have minor muscle weakness, sometimes associated with muscle hypertrophy.

Fig. 8.2 shows the distribution of new mutants in a population consisting entirely of families with one boy and one girl. Although such tidiness is not realized in nature, the consequences are similar, and only

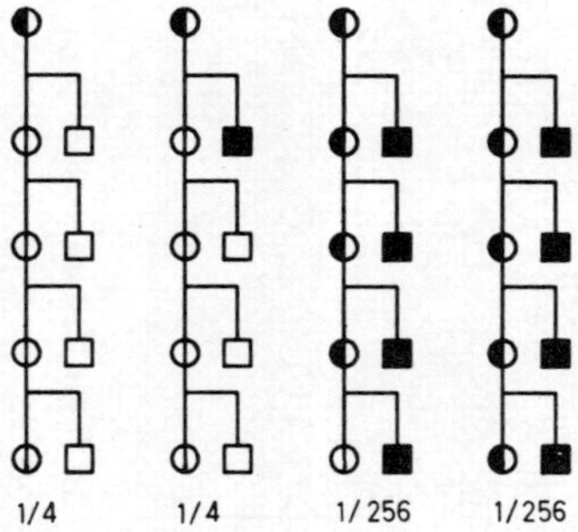

Fig. 8.2 Distribution of new lethal X-linked mutants in a population consisting entirely of families with one boy and one girl.

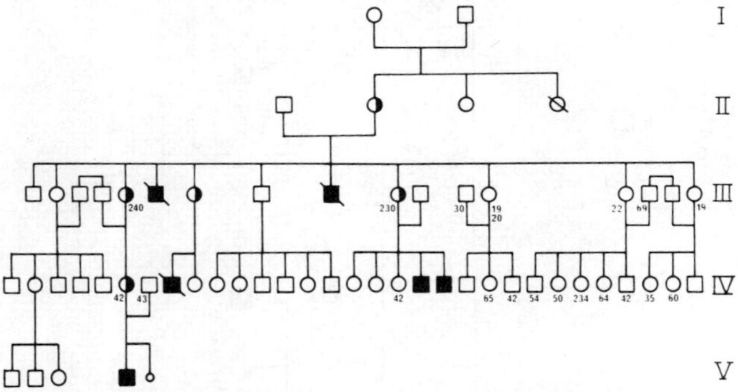

Fig. 8.3 A large family with many members with Duchenne's muscular dystrophy. The small circle (V.5) is a fetus which was sexed at sixteen weeks, and would have been aborted if male. The numbers are creatine phosphokinase levels.

a minority of cases will be preventable, most mutations leading to less than two affected cases. The proportion of girls who are carriers cannot exceed twice the incidence in boys.

Fig. 8.3 shows a large family with many members with Duchenne's muscular dystrophy, and the levels of creatine phosphokinase, which are sometimes misleading (the upper level of normals is about 60 units).

In haemophilia (Fig. 8.4) the problems are similar but the disease far less severe. Further, since the active substance, factor VIII, can be estimated both functionally and immunologically, carrier diagnosis is more precise and rapidly improving in precision. The demand for either contraception or abortion is very slight, as might be anticipated from the success of treatment and the support given by the haemophiliac societies. Most mutations seem to be in the male, since most mothers are carriers. In Duchenne's disease only about half the mothers of sporadic cases are carriers.

Some other rare disorders, of which Lesch-Nyhan's syndrome is the best known, and the most terrible, can be diagnosed *in utero* by enzymatic assay. Most are very rare.

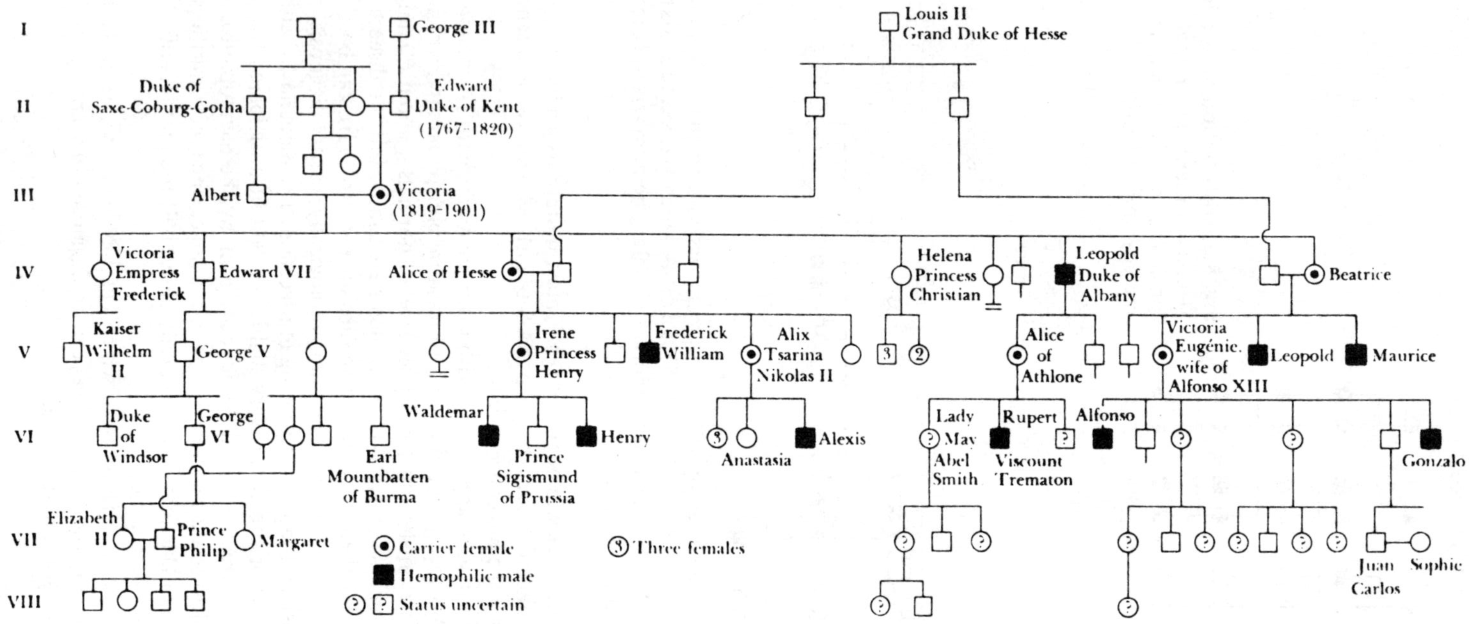

Fig. 8.4 Pedigree of Queen Victoria and her descendents, illustrating the transmission of haemophilia, an X-linked recessive trait.

8.5 Malformation

Genetic disorders are relatively rare compared to congenital malformations, that is, to malformations present at birth. Some congenital malformations are of a simple genetic nature, such as the lobster-claw anomaly of the hands, or the generalized deformities found in the three commonest trisomic syndromes (trisomy 21, 18 and 13 with incidences at birth of 1/700, 1/3000 and 1/5000).

The focal malformations, that is those malformations which are the consequence of a focal anatomical defect, almost all have a low recurrence risk, usually about 1 in 30, with a tendency to 'breed true'. Parents often ask advice, and as they are usually relieved to know the risks of recurrence are so low, it is unlikely that the event, or the advice, has much effect on subsequent fertility if the condition is either fatal or mild.

Unfortunately, we must assume that these predispositions are due to some people deviating from the mean in one direction: a vast eugenic campaign, which would clearly have to cause more non-births than the standing population, would eventually succeed in making any malformation substantially rarer, but it is not clear what else it would do since to 'shift' the mean in developmental timing one way, while reducing one hazard, might be expected to increase some other.

Particular care should be taken not to confuse specific genetic markers which are associated with disease with these markers being 'unhealthy'. Most patients with ankylosing spondylitis have the HLA-B27 allele, and most criminals carry a Y-chromosome. In neither case can the associated determinant be regarded as unreservedly undesirable. As yet there is no indication for a bad prognosis based on normal variants. So far as B27 goes the advantages of carrying this allele are unknown; there must be some advantages or it would not be so common.

Artificial selection for disorders not of a simple genetic nature is unlikely to yield benefits for the future commensurate with the disturbance to the present.

8.6 Other conditions

Generalized disorders known to be influenced greatly by sex, by race, and by country, such as coronary artery disease or diabetes, are similarly unlikely to be influenced much by selection. Nutritional habits have changed greatly within a century, and the artifical storage of foods, and the increasing distances between the origin and destination of foodstuffs, have inverted the selective advantage of the fat. All the organs of the body have to share a single blood bath, and while the disadvantages of some organs being exposed to high levels of cholesterol, or uric acid, or glucose, is clear, the optimal levels for others is not, and some caution is needed before implying that those with a tendency to atheroma, diabetes, or gout are necessarily purveyors of a predisposition for which their descendents will not thank them.

9 Mutation and disease

Man, like other species, has evolved by an ability to survive his environments, and to accommodate to changes in these environments. This has been achieved by maintaining a sufficient genetic variability for the action of selection, and this variability is necessarily based on mutation.

9.1 Mutation rate

The rate of mutation can be optimal in an evolutionary system, being bounded by zero levels, which allow no replenishment of variability after its exhaustion by selection and by chance, and by high levels which give inadequate opportunity for the clearance of mutants, a procedure which is costly since most mutants are disadvantageous, and the cost of clearing *n* mutants is of the order of *n* individuals who must be removed by death, or otherwise prevented from reproducing. This 'cost' varies from *n* for lethal dominants, either genic or chromosomal, to figures exceeding *n* for the inefficient selection against alleles with minor effects in the heterozygote. However, the numerical value of *n* distorts the impact, since as *n* increases the 'cost' to the individual casualty declines due to the consequences of affliction being mild or late.

We do not know whether our present mutation rate was ever optimal. There are many mechanisms for protecting the germ cells, and repairing DNA damage, but these may not be sufficient to have reduced the error rate to its optimal value over the past million years or so.

We do know that mutations, both recent and remote, are a cause of great distress to individuals and to potential parents, and that any increase in these mutation rates will lead to an increase in private tragedy. The tolerance of the species to an increase in mutation rate is not known, but this public danger involves grave responsibilities. The consequences of mismanagement will not be evident for many generations, and the capacity for clearance may be easily exceeded, especially as the variation in numbers of children per woman, per man, and per couple is likely to continue to decline.

Artificial selection can allow death to precede birth, and reduce the consequences to those already born, but it is necessarily restricted to conditions which are capable of diagnosis by studying cultured amniotic cells or by chromosomal examination. Various malformations may also be diagnosed *in utero* by other means, but most of these cannot be termed genetic disorders in a meaningful way. Such procedures can make little impact on the scars we may impose on our descendants in our quest for cheap power or cheap defence.

Unfortunately, very little is known of the mutation rate in man, and even in the mouse the available data are largely restricted to changes in hereditary units whose mechanism is unknown, and which may or may not be genes in the accepted sense of the word. Most of these disorders are dominant, many affecting coat colour, and few have any analogy with human dominant disease.

However, the dose necessary to double the mutation rate in the mouse is known, and is largely independent of the rate at which the dose is delivered. That is, there is no threshold. If the dominant disorders in man are assumed equally prone to a doubling in their mutation rate, then we can estimate the proportional increase in dominant disease given the proportional increase in radiation.

Unfortunately, this will not help much since it gives little information about the recessive disorders which provide the bulk of human genetic tragedy, and no increase in their frequency will be evident for many generations. This is because a unique mutant of a recessive nature will be harmless until it is paired against itself, and this can only happen through a conception by two gametes whose common ancestor was the source of the mutation.

9.2 Nature of mutation

There will also be other consequences of disrupting the hereditary information, in that the specification of development will necessarily be vaguer, even if there is no catastrophic omission, or contradiction, in the developmental message. Even less is known of these consequences. The basic consequences of mutation are shown in Fig. 9.1. in which the squares are individuals, and the arrows gametes conveying genetic units (genes). The mutational event leads to a mutant genetic unit, or mutation (●), and this may lead to a mutant phenotype, or mutant (■).

The large rectangle above, marked E, is the environment, which contains atoms, some of which are mutable (black), and X-rays. On incorporation into living matter these mutable atoms may lead to a change or mutation in a genetic molecule either by changing their chemical nature, or by releasing energy. Similar energy arriving from without the environment, or released by the mutable atoms in the environment, might lead to the mutation of a genetic units.

9.3 Consequences of mutation

On a population scale the various consequences of a sudden mutational exposure are shown in Fig. 9.2. Nothing is known of the relative magni-

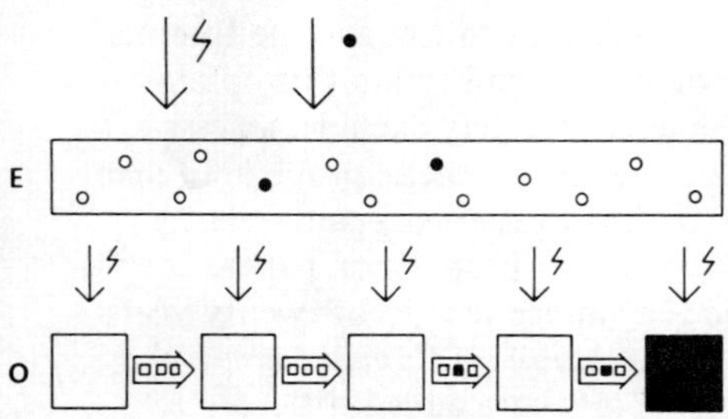

Fig. 9.1 The relationship between the environment (E) of mutable atoms and ionizing radiation, mutated genetic units and mutant organisms (O). Successive generations of organisms are represented by squares and the intervening gametes by arrows.

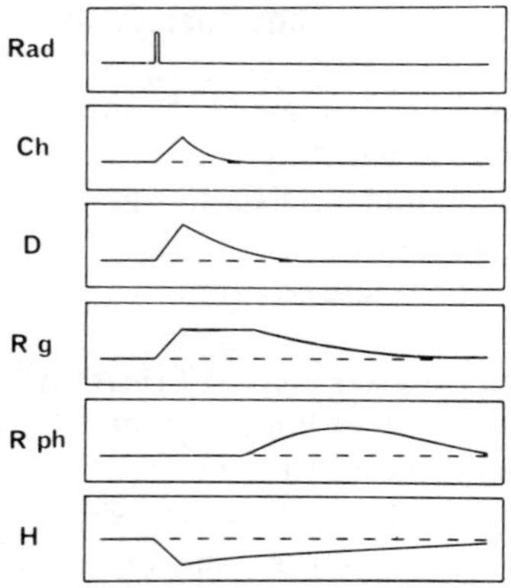

Fig. 9.2 The response to a mutational event on genetic units and phenotypes. Ch – chromosomes; D – dominant units; Rg – recessive units; Rph – recessive phenotypes; H – health. The time scale is distorted, the effect of chromosomal and dominant disease being maximized after one generation. However, the phenotypic effects of successive mutations is delayed and maximized only after centuries or millenia.

tude of these effects. Nor is there any justification for regarding the neoplastic risk as less than the mutational risk, although, since it cannot be exported to the future, it is more obvious. Furthermore, while no single somatic mutation can lead to more than one death, this is not true of genetic mutations, which are by nature not restricted to the body of those in whom they arise.

It seems unlikely that more than a minority of mutational events are due to radiation, so that a doubling of the proportion of mutable atoms, or of the amount of radiation, should lead to a far smaller effect than a doubling of the mutant genetic units.

If we could enumerate the many types of genetic units, and knew their natural mutation rate, we might be able to make some estimate of what would happen if it were doubled.

However, in man we have very limited data, in that when the mutation rate is sufficient for its rate to be measured, we are ignorant of the unit involved (excepting in the chromosomal disorders). We have limited help from mice since they do not show the same mutations as man, and are not amenable to similar forms of scrutiny. It seems likely that the limits of exposure recommended as safe for the living will be tolerable for the unborn, but the evidence for this assertion is merely assertive, and in part relates to the political realities in which safety for the living is seen as a problem to be resolved between managers and unions within the framework bounded by the constraints of engineering and the distress of unemployment. The unborn do not feature much, and the International Committee of Radiation Protection seems to limit its opinion to one or two generations, and to exclude, or largely exclude, recessive disease. The view that recessives are the major genetic problem of man is hardly unsupported by those with direct experience, although dominants seem to be regarded as the graver problem on paper.

The matter is complicated, and the vintage reports of twenty years ago deserve reading. (See '*Effect of Radiation on Human Heredity*' W.H.O. Geneva 1957). Subsequent reports have added little.

The standard of supervision in the nuclear industries in Britain is sufficient to make any increase in cancer risk so slight that its existence is in doubt, and it is certainly less than that induced by cigarettes in workers who expose themselves. Nor is there any evidence of any mutational event to workers or their children. Those exposed at Hiroshima and Nagasaki are similarly free from any evident increase in the incidence, or variety, of genetic disease. However, this gives no information on recessive mutations which will not be evident until sufficient generations have elapsed for adequate proportions of children to have an exposed common ancestor through both parents.

There is evidence that radiation can predispose to chromosomal disorders in children, and this is hardly surprising. There are adequate grounds for always protecting the gonads during X-ray, and for regarding failure to do so, when it is feasible, as a form of malpractice. Chemical mutagens may well be imposing a graver hazard than radiation; however the extent of this is also largely unknown, as are the implications of mutations arising in bacteria when chemicals have been exposed to mammalian enzymes (the Ames test).

At present tests for mutagenesis are being used as a cheap way of looking for carcinogenesis, which can only be tested on intact mammals. A test which is mutagenic and not carcinogenic is, of course, only marginally less hazardous than one which is also carcinogenic.

9.4 Mutational disease due to medicine

The destruction of tissues by radiation is known to convey risks of mutation, but the age distribution of cancer, the severity of prognosis after such exposures, and the availability of sterilization, make this a minor contribution.

Recently, the successful whole-body treatment of leukaemia in childhood, the good prognoses after transplantation when associated with antimetabolite treatment, and the use of antimetabolites for various disorders of defense, including nephrosis, have liberated large numbers of fertile individuals whose experiences would seem to make them most unfitted to reproduce, but which often make them want to live fully and not die childless.

The magnitude of this problem is potentially serious, and it may well exceed that of all other medical activities in leading to future problems. The only solution, since treatment cannot be withheld, and nor can sterilization be imposed, is the diversion of research resources to preventive measures, or to alternative forms of therapy.

9.5 Mutation and war

The mutational potential of explosive nuclear bombs appears, on rough estimates, to be such that delayed genetic deaths would be of the same order as instant deaths, a conclusion reached by Haldane and based on early bombs. Muller reached similar conclusions.

It seems clear that the total fatalities which will result from testing bombs has already comfortably exceeded the number of fatalities from the two bombs which have been used. The less explosive or nucleon bombs are likely to have a far higher remote to instant casualty ratio.

10 Some further topics

10.1 Abortion

Abortion is the act of giving birth before 28 weeks, and may be spontaneous or induced. The product is an abortus, and its plural abortuses. Miscarriage is sometimes used to imply spontaneity, but in medical use the word abortion is used for both without implication unless qualified.

The spontaneous abortion rate is at least 15% per conception; higher rates have been suggested, but the limited data from artificial insemination make rates of above 20% unlikely. In mammals, such as the cow, in which good data are available, the rates are similar, and there is no reason to expect a higher incidence in man. Over half these abortuses are due to chromosomal error, at least 4% of conceptions having an extra chromosome, a missing sex chromosome, or an extra set following double fertilization (triploidy). Most of these occur in the first three months. Spontaneous abortions after sixteen weeks are rare. Abortion is normally induced by suction in early pregnancy (before eight weeks), and by inducing labour with prostaglandin in later pregnancy. The abortion of one of a twin pair is not compatible with the survival of the other.

10.2 Artificial insemination

The demand for this is unlikely to compete with natural methods, and consequently there seems little effect on selection. The chance of unknowing inbreeding is slight, and the consequences of such events not necessarily serious.

10.3 Association

Association of some pair of phenotypes (such as duodenal ulcers and blood group O in the ABO system), or sunburn and red hair, merely

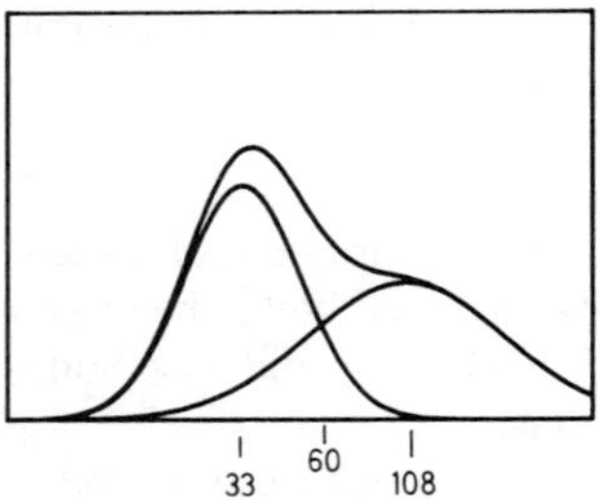

Fig. 10.1 The distributions of controls (left) and carriers (right) for creatine phosphokinase (CPK), in a population consisting of equal numbers of each. Wherever a line is drawn errors will be made. The geometric means are shown. Logarithmic scale.

means some causal relationship at the functional level, or 'neighbourliness' on the metabolic map. It does not imply linkage which is a 'structural' relationship, although very closely linked loci may be indistinguishable in their effects from linkage.

Association only implies linkage when the phenotypes associated are clearly segregating mendelian markers. Associations between other phenotypes, or between such a phenotype and a marker, does not imply linkage.

10.4 Carrier detection

A relation of someone with a genetic disorder may wish to know if they are a carrier, or, if this is impossible, of the chances that they are a carrier. The word carrier is used in several different senses, which we will consider separately.

Translocation carrier. Some individuals, about two per thousand, have their autosomes arranged in an unpaired, or unbalanced, form, and these unbalanced forms have often existed for many generations only to be revealed by a chromosomal examination for other reasons, or through an unbalanced and disabled child, or through a series of abortions. Detection is simple and reliable: it is normal practice to continue along unbalanced members of a family tree: all those with balanced chromosomes can be reassured, as can all those relatives separated from carriers by a normal member. The larger the translocation the smaller the chance of unbalanced forms surviving early pregnancy.

Dominant carrier. This is a loose terminology for dominant disorders not yet expressed. For example, the common form of the rare disorder of hereditary cancer of the colon is the consequence of thousands of small polyps which usually develop in late childhood and usually go malignant within thirty years. Diagnosis is by direct examination and treatment by excision.

In Huntingdon's chorea, a more common and more distressing problem, children of affected persons do not know if they are going to be affected, in which case they may convey the disorder to their children. They can only be reassured with reasonable (ten-to-one) assurance when they are too old to reproduce. In dominants, the carrier detection problem is limited to attempts to make the diagnosis by *signs*, that is manifestations evident to clinicians or biochemists, before it is evident by *symptoms*, that is manifestations obvious to the patient.

Recessive disorders. As most recessives (85% in Britain) are the first child in a sibship, and all could be prevented by the non-marriage of carriers, and some could be eliminated after fetal diagnosis, carrier detection has been proposed, and, where technically possible, as in sickle cell disease in Negroes and Tay-Sachs disease in Jews (especially Askanazi Jews), has been carried out. In practice the reassortment of those contemplating marriage does not work, the only effect being to lead to a 'taint', as a result of which 'rejects' tend to marry each other, and this makes matters worse. It seems that 'blood' is so basic a substance

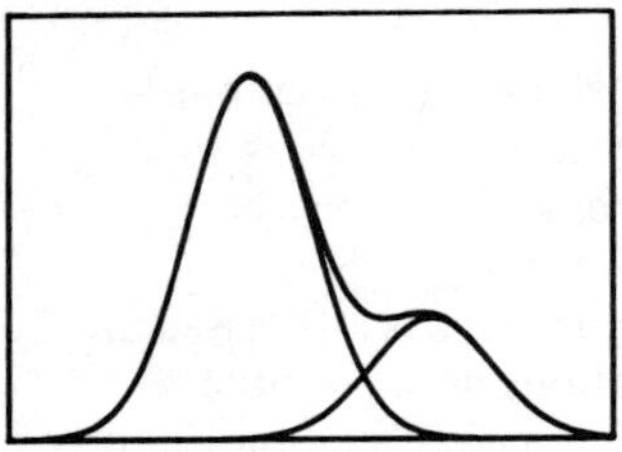

Fig. 10.2 A similar figure to 10.1 in which the populations of controls and carriers bear the ratio of 3:1 (as would be the case in the granddaughters of carriers). The risks of the two types of error are clearly different, and a different threshold is needed. When the population bears a ratio of more than 20:1 the test is too unreliable for use.

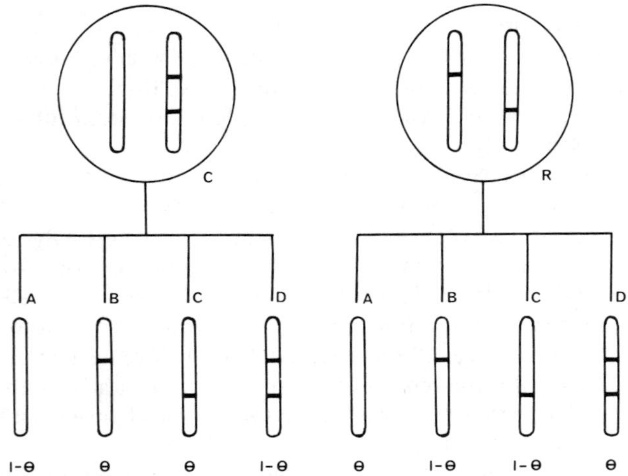

Fig. 10.3 The assortment of two chromosomes into four gametes (this is realized only in the male, but the consequences are the same in the female). On the left the distinctive alleles, marked as bars, are on the same chromosome (in coupling); on the right they are in repulsion. The proportions after assortment are as shown, and define θ, the recombination fraction. The estimation of θ is simple only if it is known whether the alleles are, or are not, in coupling.

that in no population can more than a minority fully accept that 'carriers' make ideal spouses for non-carriers. See also Cousin Marriage (p. 63, Section 10.5).

X-linked disorders. Severe X-linked disorders are frequently the result of a recent mutation: if the mutation is from the male, then all mothers of affected persons are carriers, as seems to be the case in Haemophilia (incidence 1 in 10 000 boys) and Lesch Nyhan's disease (incidence 1 in 500 000 boys). In Duchenne's disease (1 in 5000 boys), in which reproduction is unknown, many mothers are not carriers, and some, if not all, mutational events must be from the female.

Tests may be divided into 'soup' tests, where the average results of large numbers, usually millions, of cells are studied, and 'cell' tests, in which individual cells, or the progeny of individual cells, are studied. Due to Lyonisation (see p. 18) cell tests are definitive of carrier status.

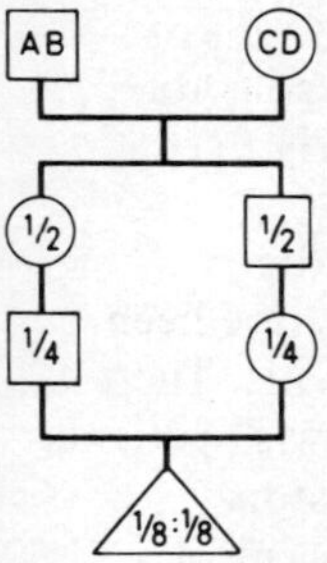

Fig. 10.4 A cousin marriage and the chance of each locus A, B, C and D being present in each descendant, and in the product of a cousin marriage. This child has a chance of 1/8 of receiving a locus from one parent and 1/8 of receiving the same locus from the other, or a probability of receiving both of 1/64. There are four ways in which this could happen since there are four loci, so the chance of identity by descent is 4/64 or 1/16. This is known as the inbreeding coefficient.

'Soup' tests, as in the estimation of creatine phosphokinase, which leaks faster from diseased than from normal muscle, and which is found in the blood, are rarely definitive, due to overlap of the levels in carriers and others (see Figs. 10.2 and 10.3).

This is the same sort of problem as defining the sex of adults given height only. Most five foot adults are women, and most six foot adults are men, but the considerable overlap makes it an unreliable method. In this case the origin of the adults will greatly influence information conveyed by height. In extreme cases, as in a nunnery or a barracks, height would be of little value in deciding sex.

In practice, in both muscular dystrophy and haemophilia, the tests are not good enough to stand on their own, and only of use in advising close relatives. They should not be done on unrelated individuals, since the majority of those with 'positive' results will be normal. Around one woman in 3000 is a carrier for Duchenne's disease, and about 1 in 5000 for haemophilia.

10.5 Cousin marriage

Advice is sometimes sought on this by cousins about to marry. In some groups, as in Muslim immigrants to Britain from Pakistan, cousin marriage is the norm, and, as expected on both theoretical grounds, and on the pragmatic grounds that it has continued for many centuries, no gross excess morbidity seems to result.

Recessive disorders are the consequence of marriages involving carriers for the same defect, and, in view of what is known of the numbers of our ancestors, and of the mutation rate, most carriers are likely to be the descendants of the same mutant. Cousins are certainly more likely than others to have children with these disorders, most of which are severe, untreatable, and afflict the brain, the eye, and the ear. However, the total number of casualties is not determined by the mating structure, which merely influences allocation.

From a practical point of view the number of recessive lethals per gamete is probably not much more than 0.5. If we study all the ways in which the four loci A, B, C and D could get from a couple to their grandchildren who were first cousins (Fig. 10.4), then a couple so related would have a 1 in 4 chance of being identical by this descent for each locus, and a 1 in 16 chance of a child being homozygous for an allele derived from a grandparent. The risk is approximately $n/16$ where n is the number of severe recessive alleles per gamete.

In practice the risk is probably 1 in 20 – 1 in 30 or so in the British population, and possibly higher in some groups, who have been through narrow 'population bottlenecks' in the last few millenia. There are no relevant tests for the aboriginal British population. Jewish patients could be tested for Tay-Sachs disease, and those whose ancestors were exposed to malaria for many centuries, and this includes most Asians, all Negroes, most Italians and Greeks, and all Cypriots, could be tested for sickle cell status (which is within the competence of many laboratories) and for thalassaemia (which is difficult).

10.6 Dermoid

Isolated masses which contain several recognizable tissues, and arise especially in the ovary and testis. These are usually due to meiotic errors, and genetically resemble the consequences of self-fertilization. The commonest form is the ovarian dermoid which develops in childhood and may contain teeth and hair. Teratomas are malignant dermoids.

10.7 Genetic engineering

The term should mean the artificial interchange of hereditary material between species. Species are, by definitions, groups with common ancestors whose hereditary material is normally only interchanged within the species, and most species have procedures ensuring a 'scorched earth' cytoplasmic policy, by which the generation gap is almost exclusively traversed by species-specific DNA with very little contamination from even RNA produced before gametogenesis.

Artificial hybrids, as the mule and many agricultural plants, can be produced by normal methods of reproduction. More extreme hybrids are only possible by inserting nuclei into alien cytoplasm by allowing the cells to fuse artificially, or to incorporate chromosomes or parts of chromosomes.

Genetic engineering usually implies that alien hereditary material is not merely acquired, but incorporated into the genetic material of the host. When the host is a bacterium this normally means inserting an alien segment into the ring chromosome, after which this segment will reproduce itself, and be capable of metabolic activity.

No example is yet known of a protein whose genetic determinants are not consistent with lineal evolution, so the opportunities for the natural transfer of genetically active material between species seem limited. The use of bacteria to incorporate bits of mammalian DNA is, in principle, capable of transmitting between individuals DNA capable of inducing tumours, or liberating virus determinants.

10.8 Incest

Incest is illegal, and there are such obvious evolutionary advantages in encouraging dispersion that the widespread and inborn aversion to this practice is not surprising. The purely genetic consequences of relaxation seem limited, since the consequences to society, and the physical level of disease, are those of allocation of children crippled by recessive disorders. The total numbers of such children will not be increased, but they will be concentrated in such marriages. Due to the strong social aversion, incest is usually associated with intellectual and other limitations, and the children of such unions are often handicapped for those reasons.

10.9 Intermarriage

The division of man into overlapping, but largely distinguishable, racial groups results from most individuals, by necessity or choice, acquiring a spouse of the same race (and, incidentally, as is sometimes forgotten by those confounding racism and sexism, of different sex). Even where opportunity exists, racial assortment on marriage is the usual, and presumably, on average, the preferred, mode. Further, this preference extends within the group to assortment for height, hair colour, eye colour, and even such minor details as elbow width (presumably due to secondary association with height). There are also the obvious marital associations by intellect and by specific aptitude (e.g. musical ability), which are in part related to opportunities for meeting. Studies of temperament show limited selection preference.

There is no substantial medical evidence against intermarriage: certain theoretical problems involving pregnancy when the male is far larger than the female, or facial structure in children when the parents have very different jaw sizes, do not seem practical issues. Fetal size seems largely determined by the mother, as in the mule and the Shire-Shetland cross. Some cases of dental disproportion are reported from South Africa.

The only genetical evidence is favourable, both on the vague grounds of 'hybrid vigour', which seems a general feature of animals and plants, and of the lower risk of a child with recessive disease. Sickle cell disease and fibrocystic disease are most unlikely to arise in Caucasian-Negro children, and Tay-Sachs disease is a most unlikely consequence of a Jewish-Gentile marriage.

The distinct problems posed by blood from immigrants having a higher incidence of infectivity from hepatitis, and therefore more appropriately diverted to indirect use than to direct transfusion, are nothing to do with genetics.

10.10 Legislation on sexual activities

Since aberrant sexual activities are, by nature, sterile, those who wish to pursue them will usually make a below-average contribution to the next generation, and, for simple reasons, no society can survive if sterile minority pursuits become too common. It is difficult to see how legislation will affect the next generation. Social or legal pressures to conform and reproduce are, if anything, likely to increase such problems in the next generation.

10.11 Linkage
When loci are on the same chromosome they are said to be 'linked' and, when close together, this may be manifest through the alleles in relatives tending to be transmitted together from parent to child. Linkage is normally measured in units of recombination, a centimorgan being a length of chromosome such that there is a 1% chance that the loci at its ends will be separated at gamete formation. Due to the mechanism of crossing over, loci at the extreme ends of a chromosome will have a 50% chance of crossing over. Loci in different chromosomes will also assort independently (see Fig. 10.2).

10.12 Lod
An acronym for the logarithm of the odds ratio. If there are two coins, one double-headed, and one is tossed six times and is always heads, the 'odds ratio' for the coin being the double-headed one is 64:1; if expressed as a lod this is log(64) or 1.81. Lods have the advantage of additivity, and are convenient in adding data.

10.13 Parental age
Chromosomal aberrations, and also some rare dominants, are commoner in the children of older parents, and the age distribution of parents can be influenced by such public measures as the control of mortgage rates. Unfortunately, while the genetic hazards are less with the younger parent, the social environment within which the child must be reared is usually worse, and no overall benefit is likely from encouraging earlier births or shorter intervals between births.

10.14 Paternity
Paternity can be excluded, but never confirmed, by the finding of genetic factors in a child absent from both the mother, whose maternity is assumed, and the putative father. Very high technical standards are necessary, and exclusion is best based on more than one system. There is relatively little enthusiasm for these tests in Britain.

The incidence of non-paternity in children conceived after marriage is about 1% in the British born population. It is probably lower in Asian immigrants. It is common to find discrepancies in a child conceived before marriage, and caution is then necessary before giving a bad prognosis for a recessive disorder. These data are based on personal experience from blood grouping of families introduced by a sick child, and are doubtless biassed.

10.15 Pedigree drawings
Pedigrees are a convenient way of representing relationships, and are usually understood by patients without difficulty. The classical 'Mars and Venus' symbols (♂ and ♀) are now largely replaced by squares and circles, which have the advantages and disadvantages of making the sex

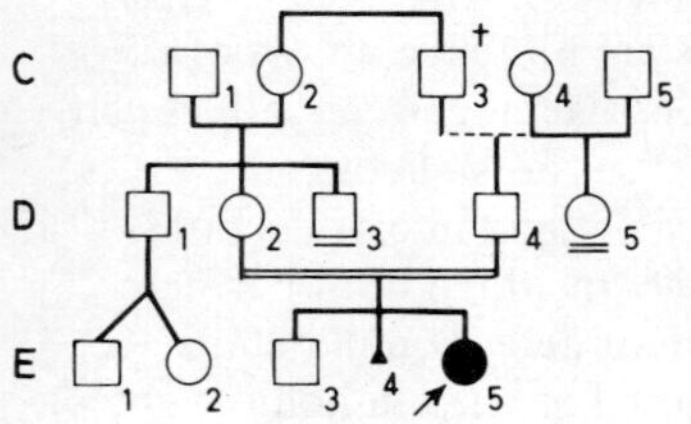

Fig. 10.5 Pedigree showing an affected girl (E5) whose parents (D2 and D4) are first cousins whose mother (D2) had one miscarriage (E4), of unknown sex, whose only cousins are twins (E1 and E2), and whose father (D4) was born outside marriage. His father (C3) is dead and his step-daughter (D5) unmarried.

of the individual more conspicuous. This is an advantage in X-linked conditions, but a distraction in most others.

Males are square or, as in life, less rounded.

Pedigrees are normally defined by Roman numerals giving the generation and Arabic numerals defining the number from the left. Bold capitals in alphabetical order, may be used, and the generation including the propositus can usefully be called P. Sibs are drawn eldest to the left. The connecting lines between generations, which represent the genetic paths, should go into the top, and come out of the bottom, of the symbols.

Various additional information may be conveyed by shading, etc. or by the use of numerals, blood group symbols etc. Dotted lines may be used to specify doubt about sib order, illegitimate unions, or doubtful paternity. A single line under an individual to denote married but childless, and a double line to denote unmarried, is useful. Death can be represented by a small cross. Miscarriage or abortion by a half size symbol. Unknown sex can be represented by a diamond or triangle.

An arrow is often used to point to the member by whom the pedigree was introduced.

10.16 Public selection

By public selection we consider the effects on the incidence of future disorders, of body or of mind, of population policies, by which is meant policies directed at the whole population, or a whole subset of a population. This is a wide issue, and one rarely considered, although liable to have a far wider impact on the genetic constitution of the next generation than the consequences of private selection. The bulk of selection for health, ability, or disability takes place in the population at large, rather than in that minority distinguished by inherited disease.

In principle if genetic advice were offered to those who did not request it some reduction in incidence would follow, since some persons not aware that their potential children were at risk could be enlightened. However, even this simple approach presupposes the availability of accurate registers of disease which were not confidential between relatives, or of reliable tests of carrier status. Registers are difficult, since accurate information is difficult to acquire unless confidentiality is assured, and the removal of those erroneously enrolled can be difficult (particularly if computers are involved), since it is difficult to control or account for old listings. At present there are no reliable tests of carrier status for diseases common in Britain. Haemoglobinopathies can be defined in immigrants, but little action is yet possible outside very specialised centres. Tay-Sachs disease can be defined in carriers but is too rare to justify screening outside marriages in which both partners are Jewish. Cystic fibrosis is not yet capable of detection in heterozygotes (at present it cannot even be diagnosed *in vitro* in homozygotes).

Little is known of the effect of such advice. It is unlikely to influence marital selection, and is likely to be most effective in severe disorders capable of reliable antenatal diagnosis. In this case the test could be left until pregnancy was established.

The proportion who might benefit from advice, but do not accept it, could be increased by reliable and balanced journalism and educational programmes. Several television programmes have been well researched, but written articles often lack balance or data.

10.17 Screening

A screen is a device used by gardeners to sift out stones from earth, and screening refers to a procedure which divides populations into two classes. In medicine it relates to procedures done on potentially healthy persons to designate minorities in need of special care.

A 'positive' is someone picked up by such a screen, and a 'negative' someone not picked up.

The nomenclature is as in Fig. 10.5.

		TEST	
		Negative	Positive
DISEASE	Absent	a	b
	Present	c	d

It is clear that there are four possibilities:

a = true negative b = false positive
d = true positive c = false negative

As the implications of false negatives and false positives differ profoundly, any screening policy must be based on the numbers a, b, c and d.

In genetic disease population screening includes rhesus blood grouping and antenatal testing on all pregnancies, neonatal blood tests for phenylketonuria, (incidence 1:10 000 in non-immigrants) and sometimes other disorders on heel-pricks (the disease is not manifest in cord blood).

Screening for chromosomal variants at birth is sometimes done as a research exercise, but is of limited value since the psychological impact of finding minor or poorly understood abnormalities exceeds the benefits possible from applying knowledge.

Screening for fetal disease is possible, and the availability of a service for women over forty is recommended in Britain. When facilities are adequate to ensure safe amniocentesis and reliable reporting, the age could be reduced with advantage. Most children with mongolism are born to women under thirty-five.

10.18 Spina bifida

It is sometimes termed a multifactorial disease, and, like all disorders adequately studied, has the substantial incidence in sibs considered above. Although no more 'genetic' than most other disorders of either fetal or adult life, the possibility of diagnosis by amniocentesis, and the administrative integration of amiocentesis with genetic departments, has led to this disorder being considered as almost an honorary genetic disease.

It is detected at amniocentesis by the 'weeping' of fetal proteins through the thin membranes which replace normal skin. Alpha-feto-protein, which replaces albumin in the fetal blood, and is almost absent from adult blood, is usually tested for. The discrimination can be adequate for accurate diagnosis. (Fig. 10.6)

This leakage is reflected by maternal blood levels, allowing a proportion of cases to be detected by testing blood. At present this would

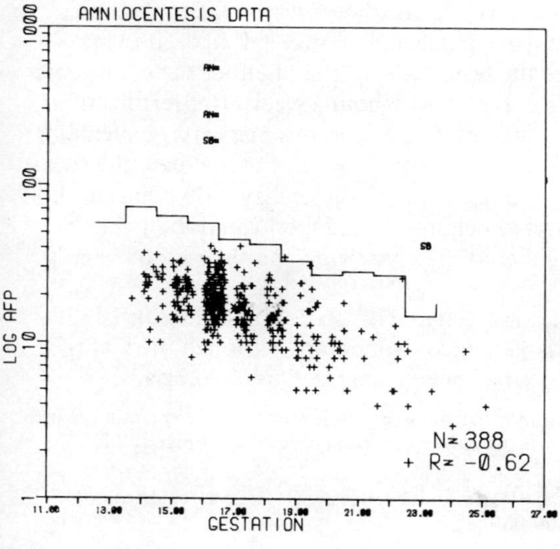

Fig. 10.6 Alpha feto-protein levels in amniotic fluids at Birmingham Maternity Hospital (by kind permission of Mrs. J. M. Crawley), showing high levels in two Anecephalic (AN) and two Spina Bifida (SB) fetuses subsequently aborted.

involve in 4000 pregnancies* selecting the upper 5% on level (200 pregnancies), defining the twins (the commonest cause of high levels) by ultrasound, and then doing amniocentesis on the rest. This would yield the elimination by abortion of about five of the eight spina bifidas expected, at a cost of one or two normals from the amniocentesis itself or from diagnostic error. The feasibility of the procedure depends on a sufficient standard of ultrasound to define twinning, and such standards are not yet widespread.

10.19 Stillbirth

A Stillbirth is a baby born after twenty-eight weeks (before which it is an abortion) and born dead. The exact definition of death at birth is difficult and is often decided on theological or humane grounds (stillbirths are less traumatic psychologically and financially to the parents, but escape the possible advantages of a full baptism). The causes of stillbirth are often mechanical, and due to delivery, or due to exhaustion following a very protracted delivery; only rarely is a genetical disorder present. Rhesus disease, once a common cause, is now rare. The 'fetal achondroplasia' and allied conditions are not usually genetic. Incidence studies on stillbirth, unspecified by cause, are of no value in the study of genetic disease or mutation.

10.20 Twins

In Britain about one pregnancy in eighty bears twins, so that about one baby in forty is a twin. Most twins are simple multiple pregnancies deriving from double ovulation, and are known as dizygotic twins. Apart from any untoward consequences of over-crowding in the womb, or difficulty in getting out, or of inadequate natural feeding, such twins are as different as other sibs. The placentae may be separate or apparently fused although the circulations are always distinct, and the inner and outer fetal membranes (the amnia and choria) enfold each twin.

A minority of twins are the result of splitting of either the embryonic cell mass before it embeds in the uterus about a week after fertilization, or of splitting of the embryonic nodule within this mass after embedding, or of splitting of the embryonic disc even later. In the first case the twins embed separately and develop in the same way as dizygotic twins; in the latter two cases, known as monochorionic and monoamniotic twins; respectively, the placentae are intimately related and their blood vessels communicate freely.

The basis of twin studies was stated with clarity by Galton in 1883, but appears to have been largely unknown before, although the identical sex of monoamniotic fetuses had been recognized by theologians.

"The reader will easily understand that the word 'twins' is a vague expression, which covers two very dissimilar events – the one corresponding to the progeny of

*A large maternity hospital handles 4000 deliveries a year, a large region about 40 000 and the U.K. about 600 000.

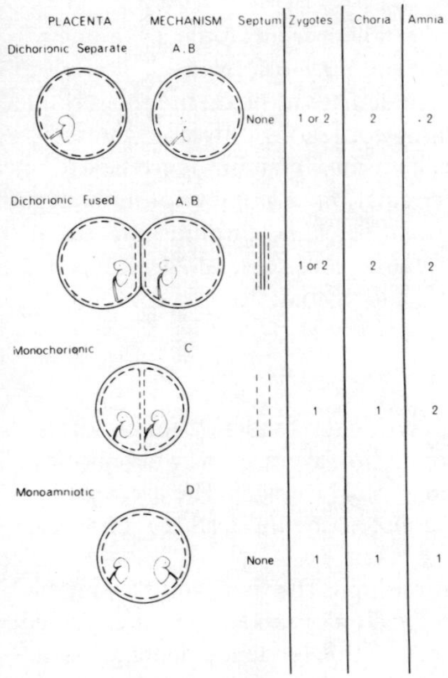

Fig. 10.7 The relationship of zygotes, membranes, and types of twinning.

animals that usually bear more than one at birth, each of the progeny being derived from a separate ovum, while the other event is due to the development of two germinal spots in the same ovum. In the latter case they are enveloped in the same membrane, and all such twins are found invariably to be of the same sex. The consequences of this is, that I find a curious discontinuity in my results. One would have expected that twins could commonly be found to possess a certain average likeness to one another; that a few would greatly exceed that average likeness, and a few would greatly fall short of it. But this is not at all the case. Extreme similarity and extreme dissimilarity between twins of the same sex are nearly as common as moderate resemblance. When the twins are a boy and a girl, they are never closely alike, in fact their origin is never due to the development of two germinal spots on the same ovum". (Galton 1883)

Zygosity may usually be determined from the sex and the blood groups, since these must be the same for monozygotic twins and may be the same for dizygotic twins. Pairs of dizygotic twins have a 50% chance of being of different sex, and on comprehensive blood grouping over 90% are distinct. Provided that both twins are healthy and well-formed, zygosity is usually evident on casual examination of the face. Placental examination is only reliable in expert hands, and can only diagnose monozygosity with certainty as the number of placentae is an unreliable guide to the number of zygotes. The relationship of zygotes, membranes, and types of twinning is shown in Fig. 10.7. The excess of similarity of monozygotic twins, compared to dizygotic twins, has been used as a simple measure of strength of heredity, especially in intellect, behaviour, insanity and disease. However, the numerical treatment is difficult, and, in view of the convincing nature of many verbal descriptions, largely superfluous.Twins seem slightly handicapped on tests of intelligence, and this is probably due to their upbringing. There seems to be a genuine shortage of distinguished monozygotic and dizygotic twins.

They are particularly at risk to brain damage at birth; monochorionic twins are liable to congenital heart disease affecting only one; this is hardly surprising as the feedback mechanisms which influence development are likely to be upset by a common circulation. A number of monozygotic pairs have now been described discordant for some chromosomal disorders which has developed in some of the zygotic cells. Several XO/XY pairs, presenting as children of different sex, and several discordant for mongolism, are now known.

10.21 Regression

A term introduced by Galton to describe the tendency for children to resemble both a parent and the population average, or, where both parents are defined, to resemble the parental 'average'. The averages may be numerical, as in height or response to mental tests, or more complicated, as in facial appearance.

This regression extends to all relationships. The 'heritability' may be defined in terms of the ratio of the regression observed to that expected in a uniform environment. The strength of this tendency depends both on the influence of the environment, and on the extent to which the environment is common to relatives.

10.22 Recessive.

Since a recessive is only manifest when both parents provide the appropriate allele the proportion of recessives is very much lower than the proportion of these alleles. If the parents are not related, and q is the allelic proportion (gene frequency), then the chance of both being of this type is q^2. As carriers may be a carrier at either of the two loci the proportion of carriers is twice the allelic proportion ($2q$, or, more exactly, $2q(1-q)$). In cystic fibrosis, q is about 1 in 40, the incidence (q^2) about 1/1600 and the carrier rate ($2q$) about 1/20. In phenylketonuria in Celts, q is about 1/100, the incidence (q^2) about 1/10 000 and the proportion of carriers ($2q$) about 1/50.

Suggestions for further reading

References have not been given in the text, since the book is intended to be basic, and a bridge of more advanced books or reviews is needed before the original articles can be appreciated.

Blood Groups and Genetic Markers

Race, R. R. and Sanger, R. (1975), *Blood Groups in Man*, 6th edition, Blackwell Scientific Publications, Oxford.

The standard work on red cells, with valuable chapters on linkage and disease associations; written in English of exceptional clarity and elegance.

Harris, H. (1970), *The Principles of Human Biochemical Genetics,* North Holland Publishing Co., London.

This is complementary to the above book, and covers the enzymatic variants, and the implications of this variation in terms of evolution and selection.

Giblett, E. R. (1969), *Genetic Markers in Human Blood,* Blackwell Scientific Publications, Oxford.

A lucid introduction to both the subjects of the above books, and of particular value in enzyme variants.

Clinical Genetics

McKusick. V. A. (1975), *Mendelian Inheritance in Man*, 4th edition, Johns Hopkins University Press, London.

The standard reference dictionary for mendelian diseases, and an essential in any hospital or university library. The book is written in computer format, and new editions produced every two years. This is the most complete bibliography.

Chromosomal Disorders

Hamerton, J. L. (1971), *Human Cytogenetics*, Vols. 1 and 2, Academic Press, New York and London.

The standard work on human chromosomal disorder with good photographs and an inclusive bibliography. Especially valuable as a source of tables containing survey data.

Swanson, C. P., Merz, T. and Young, W. J. (1967), *Cytogenetics,* Prentice Hall, New Jersey.

A good classical introduction to the chromosome and its mechanisms, covering all species, and especially drosophila. Excellent descriptions of the mechanisms of linkage and of non-disjunction.

Defence Mechanisms.

Thompson, R. A. (1974). *The Practice of Clinical Immunology*, Edward Arnold, London.

Hobart, M. J. and McConnell, Ian (1976). *The Immune System*, Blackwell Scientific Publications, Oxford.

These books both provide a sound and lucid introduction to this difficult and changing field. The former is written from the physician's viewpoint, the latter from the general biologist's. Both include excellent diagrams and detailed references.

General Human Genetics

Carter, C. O. (1977), *Human Heredity*, Pelican Books.

A lucid and balanced introduction to human variety and disease: very little on mechanism, and designed for the intelligent non-medical reader.

McKusick, V. A. (1969), *Human Genetics*, 2nd edition, Prentice Hall, New Jersey.
Fraser, F. C. and Nora, J. J. (1975), *Genetics of Man*, Lea and Febiger, Philadelphia.
Emery, A. E. H. (1975), *Elements of Medical Genetics*, 3rd edition, Churchill Livingstone, London.
These three books are all designed to introduce genetics to medical students and doctors, and they all succeed in this. Professor Emery's book is the cheapest.
Penrose, L. S. (1973), *Outline of Human Genetics*, 3rd edition, Heinemann Educational Books Ltd., London.
This book, first published in 1959, is still the best introduction due to the unusual range of both experience and ability of its author.
McKusick, V. A. and Ruddle, F. H. (1977), The Status of the Gene Map of Human Chromosomes, *Science*, **196**, pp. 390–404.
The human chromosome map, up to going to press, with an introduction to the methods by which it has been developed, and with full references.

Mathematics and Population Genetics

Cavalli-Sforza, L. L. and Bodmer, W. F. (1971), *The Genetics of Populations*, W. H. Freeman & Co, San Francisco.
A comprehensive work on the subject with extensive documentation on population variation; those unable to follow the mathematics should not be discouraged from reading the text. The descriptions of blood groups, haemoglobins, etc., are clear, and there are many diagrams.
Jacquard, A. (1970), *Structures Genetiques des Populations*, Masson et Cie, Paris.
Translation 1974, *The Genetic Structure of Populations*, Charlesworth & Charlesworth.
A very important book. The French edition makes no extensive demands on vocabulary, but requires good eyesight and a tolerance of misprints in formulae.
Crow, J. E. and Kimura, M. (1970), *An Introduction to Population Genetic Theory*, Harper and Row, London and New York.
This, which is hardly an introduction, requires considerable mathematical ability.
Falconer, D. S. (1964), *Introduction to Quantitative Genetics*, Oliver and Boyd.
A very lucid book written by a biologist with a large experience of mice and animal breeding. Haldane considered it the best introduction to the subject.

General Genetics

Watson, J. D. (1975), *Molecular Biology of the Gene*, 3rd edition, W. A. Benjamin Inc.
This book, which assumes basic knowledge of chemistry, is clearly written and the writing is supported by extensive, accurate and elegant diagrams.
Harris, H. (1970), *Cell Fusion*, Oxford University Press, Oxford.
An introduction by the discoverer of the technique of interspecies hydridization in vitro.
Cove, D. J. (1971), *Genetics*, Cambridge University Press, Cambridge.
A useful introduction in which funguses, the author's main interest, take a leading role.

Amniocentesis

Simpson, N. E., Dallaire, L., Miller, J. R., Simonovich, L., Hamerton, J. L., Miller, J., and McKeen, C. (1976), *Prenatal Diagnosis of Genetic Disease in Canada*: report of a collaborative study, *J. Can. Med. Assoc.*, **115**, 739–746.
Galjaard, H. (1976), *European Experience with Prenatal Diagnosis of Congenital Disease*: a survey of 6121 cases, *Cytogenet. Cell Genet.*, **16**, 453–467.
This very emotional field has been badly served by journalists, and some American studies and publications should be read with caution. These two series, although somewhat biassed by the exclusion of some failures, provide the best series so far published.

Historical

Whitehouse, H. L. K. (1973), *Towards an Understanding of the Mechanism of Heredity,* 3rd edition, Edward Arnold Ltd., London.
Includes an excellent historical introduction.

Hogben, L. (1933), *Nature and Nurture*, Allen and Unwin, London.
A timeless discussion of the nature–nurture issue.

Dunn, L. C., *A Short History of Genetics*, McGraw-Hill, London and New York.
Covers the development of Mendelian ideas.

Medvedev, Z. A., (trans. I. Michael Lerner), *The Rise and Fall of T. D. Lysenko,* Columbia University Press.
The definitive history of political biology in Russia.

Gedda, L. (1961), *Twins in History and Science*, Charles C. Thomas.
Mainly of historical interest

General Reviews

British Medical Bulletin, 25, 1: *New Aspects of Human Genetics*, 1969.
British Medical Bulletin, 32, 1: *Human Malformations,* 1976.
British Medical Bulletin, 32, 3: *Haemoglobin: Structure, Function and Synthesis,* 1976.
Published by the Medical Department, the British Council, London.

Readings from Scientific American: *The Chemical Basis of Life.*
Readings from Scientific American: *The Molecular Basis of Life.*
Readings from Scientific American: *Ecology, Evolution and Population Biology.*
Readings from Scientific American: *Facets of Genetics.*
Published by W. H. Freeman & Co., California.

These can all be strongly recommended, although, as in all collections, some obscure and even misleading articles are included (for example, George the Third's porphyria). Individual chapters in the Scientific American *series can be obtained as offprints. Some volumes in this series overlap in content, and European work is understated. The* British Bulletin *series is, of course, restricted to British contributions.*

Glossary

Allele The abbreviated and usual form of allelomorph, or alternative form, of a gene.

Allosome A rarely used word defining the chromosomes related to sexual dimorphism. Synonymous with sex chromosome.

Amnion The inner fetal membrane, plural amnia.

Arm Chromosomes may have one or two arms; the longer arms are conventionally placed below the centromere.

Autosome A chromosome not related to sexual dimorphism. The autosomes are numbered, starting with the largest.

Band Chromosomes may reveal bands on staining: the term band applies to both deep and light staining segments. They are numbered from the centromere.

Barr body The resting form of one of the X-chromosomes which, in some cells, in some mammals, can be visualized in the resting nucleus. In primates this body may be seen on the nuclear membrane of most mucosal cells, fibroblasts, and some other cells, though not in lymphocytes.

Centromere The apparent constriction at which the fibres by which the chromosomes are moved at mitosis are attached.

Chimaera A mythical beast compounded of varying species. In genetics an organism devised from three or more gametes.

Chorion The outer foetal membrane, plural choria.

Chromatid One of the two strands formed when a chromosome divides.

Chromosome A dye-retaining intranuclear organ which includes the genetic material; Darlington writes 'The nucleus is the chromosomes'.

Co-dominance The co-existent expression of two dominant determinants; in the ABO blood groups A and B are dominant to O, but co-dominant with each other.

Complement A series of proteins which, when primed by a cell surface reaction, associate and destroy the continuity of the surface membrane.

Congenital Evident at birth.

Deficiency The absence of part of a chromosome.

Deletion A synonym of deficiency commonly used in relation to radiation damage.

Diamniotic Having two amnia.

Dichorionic Having two choria.

Duplication The presence of duplicated parts of a chromosome.

Dominant An allele which is manifest equally in the heterozygote and one homozygote. Used casually in human genetics to cover genetic units which are manifest in the heterozygote. Also used for related disease.

Fertilization The act of cytoplasmic fusion between male and female gametes: if successful, the nuclei of both divide and their products are included within common nuclear membranes to form the nuclei of the zygote.

Fraternal Literally, brotherly, and implying twins who are as distinct as single births to the same parents. Now usually termed dizygotic.

Gamete A sperm or ovum.

Gene That segment of DNA which determines a messenger RNA, or, in the case of RNA determinants, the RNA.

Genic Related to the genes.

Gene Frequency A term used to define the allelic proportions at a locus. E.g., the ABO locus in some Caucasians contains one of three alleles, A, B, O, with proportions 0.25, 0.05 and 0.7.

Genetic Related to the genetic apparatus (mainly chromosomes and genes).

Genotype The genetic constitution.

Haplotype The contribution of one parent to a locus or set of closely linked loci. In the former case it is synonymous with allele.

Hemizygote Related to X-linked loci in the male.

Heritability A measure of strength of familiarity in a defined and uniform environment. Sometimes used casually.

Heterozygote An individual whose contribution from each parent is different for the genetic unit considered.

Homologue The other member of the pair of chromosomes, or loci.

Homozygote An individual whose contribution from each parent is the same for the genetic unit considered.

HLA Human Leucocyte Antigen. A chromosomal region at which the determinants of the HLA antigens are located. This is on the short arm of chromosome six, and, since it has a chance of about 2% per generation of a rearrangement at meiosis, probably occupies about a twenty-fifth of the short arm.

Idiogram The diagrammatic representation of a karyotype. It may be based upon measurements made of the chromosomes from several cells.

Isoenzyme A member of a class of enzymes coded for at the same locus, and usually differing by only a single amino-acid.

Karyotype A representation of chromosomes of a single cell, arranged in orderly pairs, derived either from drawing or photography. This array can typify the chromosomes of an individual or even a species.

Linkage Loci are said to be linked when they are on the same chromosome.

Linkage Disequilibrium The association of alleles as a result of their loci being closely linked.

Lyonization The 'breeding true' of the active and inactive X in mammalian somatic tissues after their differentiation. The process of condensation (rendering heterochromatic) of an X-chromosome and its permanent imprinting on all replicated products of that chromosome.

Locus The position on a chromosome of the segment of DNA which is defined as a gene. Plural loci.

Marker A hereditary quality defined at conception and determined by a single locus, such as a blood group. Also used for any distinct feature of the phenotype, including the karyotype.

Meiosis The pair of divisions, termed I and II, as a result of which each pair of chromosomes is distributed so that there is only one per gamete.

Messenger An abbreviation for messenger RNA.

Monoamniotic Having one amnion.

Monochorionic Having one chorion.

Mono-ovular Now repalced by monozygotic.

Monosomy The absence of one of a pair of chromosomes in the diploid cell; also used with the sex chromosomes, although the identity of the missing chromosome is undefined.

Monozygotic Determined by a single act of fertilization, as in identical twinning.

Monozygous A corrupted form of monozygotic.

Mosaic An organism derived from two gametes which contains two or more genetically distinct cell types due to events following fertilization. Sometimes confused with chimaerism.

Mutant The phenotype consequent on a mutation.

Mutation A change in the genetic material.

Mutation rate The proportion of new units per defined unit per generation.

Nucleolus A nuclear organelle associated with ribosome synthesis.

Nucleotide One of the units defining the genetic code by crosslinking the two DNA backbones, as rungs on a ladder.

Nucleus The central body of the cell which contains the inherited material, and which is derived by direct descent from the nuclei of the parental gametes.

Phenotype The appearance, including the appearance of microscopic sections, electrophoretic bands, blood group reactions, intelligence test results, etc., of an organism.

Phytohaemoglutinin An extract of scarlet runner beans used to initiate mitosis in lymphocytes.

Proband A synonym of propositus or proposita, mainly used in German.

Propositus The member by which a family is introduced. Feminine form proposita. Plural forms -i and -ae.

Recessive An allele which is manifest only in the presence of a similar allele; also used of a disease related to such alleles.

Recombination The consequence of crossing over as a result of which alleles at two loci on the same chromosome may be transmitted on different gametes. The proportion so transmitted is the recombination fraction for this pair of loci.

Ribosome	A body composed of ribosomal RNA and proteins through which the information from messenger RNA, which is a sequence of nucleotides, is translated to an amino-acid sequence or protein.
RNA	Ribonucleic acid.
Satellite	A knob attached by a thin strand to the short arms of chromosomes 13, 14, 15 and 21, 22 in man. Not to be confused with satellite DNA.
Satellite DNA	Originally that DNA which, on ultracentrifugation in a gradient, was distinct from the main band. Now used for any distinct form.
Segregation	A mistranslation of Mendel's 'Spaltung' meaning splitting. The assortment of chromosomes, loci, etc. as a result of meiosis and gametogenesis.
Sib	A brother or sister.
Sibship	The set of children of a couple.
Sibling	An archaic form of sib.
Sister chromatid	One of the pair of chromatids formed by chromosomal division immediately before mitosis.
Sister chromatin exchange	The exchange of material between sister chromatids, an event now revealed by special stains leading to 'harlequin chromosomes'.
Syntenic	On the same chromosome, but not necessarily within range of detection by pedigree analysis.
Telomere	The terminal unit of a chromosome.
Transition	An interchange of one pyrimidine or one purine for another.
Translocation	The interchange of material between two chromosomes. This interchange is normally reciprocal, with preservation of the original telomeres.
Transversion	An interchange between a pyrimidine and a purine.
Trisomy	The presence in a nucleus of three homologous chromosomes instead of the normal pair.
Wild type	The commonest allele in the 'wild' or 'non-laboratory' species.
X-linked	A locus present on the X-chromosome, or a disease or trait related to alleles at such a locus. Sometimes termed sex-linked.
Zygote	The product of one ovum and one sperm. The word is derived from the yoke by which pairs of cattle were attached to a plough.

Index

G = Glossary